# Bulb Crop Cultivation

# Bulb Crop Cultivation

Mahima Singh
Ashish Kumar

2026
Agri Horti Press
New Delhi-110 084

ISBN: 978-93-49690-32-5 (Hardbound)



***Published By***

**Agri Horti Press**

New Delhi-110084

E-mail: agrihortipress@gmail.com

*Digitally Printed at* : Replika Press Pvt. Ltd.

# Preface

*Bulb Crop Cultivation* offers a comprehensive exploration of the science and practice involved in the production of bulbous crops, which play a significant role in global horticulture, cuisine, traditional medicine, and the floriculture industry. This volume is designed to serve as an informative resource for students, researchers, agricultural professionals, and progressive farmers interested in enhancing their understanding of bulb crop production systems, from foundational principles to modern cultivation techniques.

Bulb crops such as onion, garlic, shallot, tulip, hyacinth, and lily are grown across diverse agro-climatic regions and hold considerable economic and cultural importance. Their cultivation requires specialized knowledge of crop physiology, soil and nutrient management, pest and disease control, post-harvest handling, and market dynamics. As global demand for these crops continues to rise—driven by their culinary, ornamental, and health-related applications—there is an urgent need for sustainable and efficient production practices.

This book aims to provide a detailed and structured overview of bulb crop cultivation, encompassing key topics such as crop selection, propagation methods, climate and soil requirements, irrigation and fertilization strategies, integrated pest and disease management, and advancements in storage and post-harvest technologies. Each chapter draws upon current research findings, agronomic practices, and case studies to highlight challenges and innovations relevant to growers and policymakers alike.

Special emphasis is placed on sustainable production approaches, including organic farming, resource-efficient technologies, and climate-resilient practices. The integration of traditional knowledge with modern agronomic innovations is also explored, acknowledging the value of both in advancing productivity and ecological stewardship.

We express our deep appreciation to the authors and contributors who have shared their expertise in bringing this volume to fruition. It is our hope that *Bulb Crop Cultivation* will serve not only as a reference guide but also as a platform for continued research, innovation, and dialogue in the field of horticultural science. By equipping readers with knowledge and best practices, this book aspires to support the development of a more sustainable and profitable bulb crop sector.

**Mahima Singh | Ashish Kumar**

# Contents

# Chapter 1

# Selection of Bulb Crops Variety for Cultivation as per Season

## 1.1 Identification of appropriate variety of bulb crop as soil & climatic conditions

Bulb crops consist of:

- garlic
- leeks
- dry-bulb
- green onions

**Garlic bulb**

- Garlic is hardy perennial member of onion family.
- Garlic is native to Central Asia but has long been naturalized in southern Europe and throughout the world.
- It differs from onion, producing number of small bulbs called as cloves rather than one large bulb.
- In this, each bulb has dozen or more cloves that are covered with thin white skin.
- Each clove is made of two modified mature leaves around an axis with vegetative growing point.
- The outer leaf is dry sheath, while base of inner leaf is thickened, making up the bulk of the clove.
- In this, larger outer clove produces best garlic.

- Garlic has flat leaves rather than round hollow leaves of onion.
- It is used as condiment and flavoring in gravies, tomato sauces, soups, stews, pickles, salads, salad dressing and breads.

## Crop Rotation and Location

- It is good to practice rotation when planting garlic.
- It is not planted where onions or member of onion family has been grown previously.
- Plant garlic in full sun and in well-drained bed with organic matter worked into it.
- It likes well-drained soil, and addition of organic matter using hardest clay.
- As garlic requires up to 8 months to mature, plant in area where it won't be disturbed.

## Soil

- Garlic grows best in full sun and well drained soil.
- It grows best on friable, loamy soils that are fertile and have organic matter.
- The soil must be kept evenly moist as dry soil will cause irregularly shaped bulbs.
- Heavy clay soils create misshaped bulbs and makes harvesting difficult.
- Adding organic matter like well-rotted manure or compost to soil on yearly basis, keeps it friable.
- Garlic bulbs will be small if soil is excessively dry and irregular in shape if soil becomes compacted.

## Climate

- Garlic cultivation requires combination of different types of climate.
- It needs cool and moist climate for bulb development and vegetative growth while for maturity climate must be warm and dry.
- It cannot tolerate extreme cold or hot conditions.
- Exposing young garlic plants to temperatures lower than 20°C for 1 or 2 months will hasten bulb formation.
- A prolonged exposure to lower temperature reduces yield of bulbs.
- Bulbs maybe produced at axil of leaves.
- A cooler growing period gives higher yield than warmer growth conditions.
- The optimal day length requirement for bulb formation is 13-14 hours for long day garlic and 10-12 hours for short day garlic.

## Leek

- ❖ It is a non-bulb forming member of onion family which is grown for its blanched stem and leaves.
- ❖ It is eaten raw alone or mixed in salads and cooked as flavouring in soups and stews.
- ❖ It requires similar climatic and soil requirements as that of onion.
- ❖ Except of its large size, it looks like green onion.
- ❖ It is cultivated for blanched stems and leaves.
- ❖ It is biennial plant, consumed as salad and used in flavouring the soups.
- ❖ It is produced on 50 to 100 acres in small tracts for roadside and local grocery sales.

**Soil**

- ❖ It requires good aerated soil with good drainage and moisture retention capacity with pH of around 6.5-7.
- ❖ It needs deep ploughing so that longer shaft can be developed.
- ❖ It can be grown on mineral soils but in heavy clay soils, harvesting should be done and removal of extra soil from roots requires more labour.

**Climate**

- ❖ Leeks are cool weather crop that grows well in temperature of around 12°C to 25°C.
- ❖ They are not suited to growing in temperatures above 12°C to 25° and if temperatures exceed, they will mature and harden rapidly, making them unsuitable to harvest.
- ❖ They are frost resistant but excess cold weather slows the growth and affect production and quality.

**Onion bulb**

- ❖ Onion is an herbaceous biennial in the family Liliaceae grown for its edible bulb.
- ❖ The stem of the plant is flattened disc at base and tubular leaves form pseudostem where their sheaths overlap.
- ❖ The leaves are either erect or oblique and there are 3–8 per plant.
- ❖ The onion plant produces pink or white flowers clustered on stalks.
- ❖ The bulbs are formed just above the flattened stem of the plant by overlapping leaves.
- ❖ The bulb is made of many layers, each corresponds to leaf.
- ❖ They are oval but shape can be variable and occur in clusters of 3–18 to a plant.
- ❖ The bulb is protected by a membrane which turns to a papery coat.
- ❖ Onion plants can reach a height of 50 cm and are grown as annuals, harvested after one growing season.
- ❖ Onion is referred to by cultivar which includes red or purple, shallots and spring or scallions.
- ❖ The bulb is edible vegetable and is used part of onion, that consumes after cooking although it can be eaten fresh.

- The stems and leaves are also edible.
- Onion crop can be raised by transplanting, planting bulbs directly in field, broadcasting or dibbling the seeds in prepared field.

**Climate and Soil**

- Onions are hardy, cool season vegetables that grow best at temperatures of 12 to 24 °C (55–75 °F), in areas with cool spring weather and drier, hotter summer weather.
- They need fertile, well-draining soil such as clay or silt loams with a pH of 5.5–6.5.
- The plants do not do well in acidic soils.
- Onions should be set out in full sun for optimum bulb development.
- Onions are biennial vegetables and if they are left in ground for another year, produces flowers and set seed.

**Onions – green**

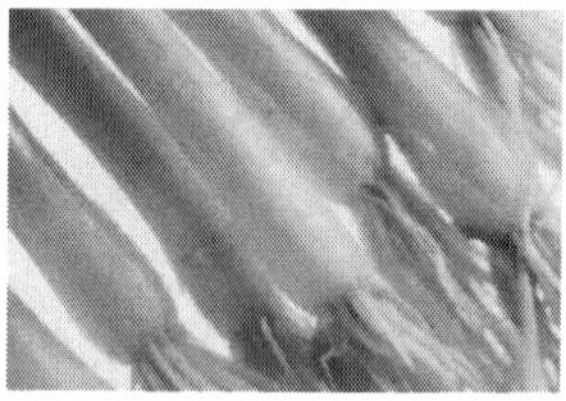

- Green onions, also known as scallions or Spring onions are immature onions which are harvested before bulbs mature.
- Seed can be purchased to grow onion varieties which are bred to be harvested as green onions.
- Green onions, like regular bulbing onions can be started from seed or sets and can be grown as transplants indoors.
- It does not need same amount of space that bulbing onions do and seed or transplants can be set out 2.5–3.8 cm apart in row to allow 5 cm between the sets.
- They can be harvested when they reach 15 cm in height or more.

**Climate**

- It grows best when temperatures is around 68 and 77 degrees Fahrenheit.
- The soil that drains quickly and doesn't remain overly wet during cool weather is good for growing.
- Sites with heavy clay soil retain moisture well during cool weather.
- Working 2 to 3 inches of compost before planting improves drainage and moisture retention.
- It can tolerate a range of soil types and can grow well in most established vegetable gardens till they receive daily sunlight.

**Soil**

- Overly acidic soil can negatively affect green onion growth.
- The plants perform better when pH is between 6.2 and 6.8, that can be determined by testing.
- Adding agricultural limestone to soil 4-5 weeks before planting can raise the pH if soil is too acidic.
- Like most onions, green onions grow shallow roots, so soil fertility must be in top 3 to 6 inches of soil for plants to access.

**Types of Onions**

**Six types of onions are grown in India.**

- Green onion are used for salads
- Mild onion are used for cooking or salad
- Pungent varieties are used as condiments for flavouring of a number of foods
- Pearl onion or small onion are used in pickles including vinegar pickles
- White and Yellow onion are used for dehydration purpose and for manufacturing onion powder
- Red onion and Yellow onion are used for general purposes for culinary uses

**Climate and Soil**

- For growing onion, ideal soil should have pH in between 6.5 to 8.
- The soil should be well aerated.
- Heavy soil should be avoided.
- Onion can grow under wide range of climatic conditions, but it cannot stand in too hot or too cold weather.
- It prefers moderate temperature in summer and in winters.
- Short days are very favorable for formation of bulbs which can be grown well at elevations of 1000 to 1300 m above MSL.
- Onion needs well drained loamy soils, rich in humus, with fairly good content of potash.
- The crop raised on sandy or loose soil produces bulbs that are deformed and during harvesting, many bulbs are broken and bruised and so they do not keep well in storage.

## 1.2 Identification of bulb crop variety as per market demand

### 1.2.1 Introduction

Marketing channel refers to the sequential arrangements of various marketing intermediaries involved in the movement of products from producers to consumers.

### 1.2.2 Marketing of Garlic

- Garlic is related to onion, leeks and shallots since they belongs to lily family.
- It has strong, heated and pungent taste.
- It grows in form of bulb with cloves each covered with papery skin.
- Garlic is important bulb crops grown used as spices or condiment in India and in world
- It is consumed in every home and used in processing industry has constant market demand.
- The garlic bulbs are easily stored for 5-6 months after harvesting which favors long marketing season.
- Garlic bulbs after reaching to assembling centers are sold to local distribution to consuming in markets.
- Market transactions of garlic usually take place using commission agents.
- Exporters either buy their requirements directly from assembling centers through local commission agents or from wholesalers in important assembling or regulated markets.
- The sale of garlic cal also takes place directly between the buyer and seller or through commission agents or brokers.
- Garlic moves from farmers to consumers through marketing channels with number of intermediaries such as wholesaler and retailer.
- There exists three channels of garlic marketing.
- Channel-I is the simplest marketing channel with no involvement of market intermediaries in trading where direct interaction is made between producer and consumer which works on minimum marketing cost.
- Channel-II involves retailer in between producer and consumers where besides relevant cost incurred by producer, substantive cost and margins occurs in part of retailer where one side reduces the share of producer of around 90% in ultimate consumer's paid price and gives no relief to customer due to reduced share of producer which finally goes to retailer resulting in expansion in paid price of consumer
- Channel III marketing is done between producer and consumer with involvement of wholesaler and retailer. In this, wholesaler and retailer search and creates demand of consumer which charges margin and cost that takes place on activities which are rendered by them, so with this, there exits difference in price received by producer and price paid by consumer.

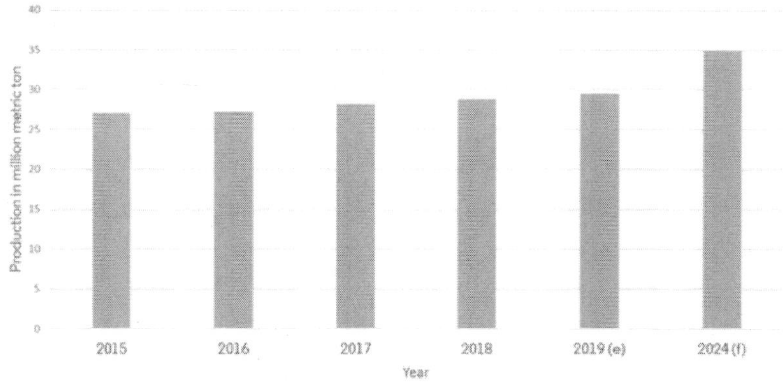


- ❖ With its use in processing and other industries for manufacturing of herbal products, pastes, medicines, pickles and curry products, the global market registers CAGR of 4.7%.
- ❖ Garlic is most important crop in many Asian countries, so Asia-Pacific holds about 91% share of global market in terms of consumption.
- ❖ Garlic contains meal which lead to reduction in cancer risk by regular intake, so such medicinal usage increases its demand from European and other countries.
- ❖ The increasing exports of garlic from India and China shows high demand for garlic globally.

**Market of Onion Bulb**

- ❖ Onions are traditionally sold loose in bulk containers.
- ❖ For marketing of onions, there exists four alternatives which involves wholesale markets, cooperatives, local retailers and roadside stands.
- ❖ In wholesale marketing, producers contract with shippers to market and ship the onions for a predetermined price.
- ❖ Marketing cooperatives use daily pooled cost and price, which spreads price fluctuations over participating producers.
- ❖ Global sales from onions was reported as US$3.5 billion where value of exported onions rises by 12.1% for exporting countries when international sales of onions valued at $3.2 billion.
- ❖ Year over year, value of globally exported onions accelerate by 9.2% from 2017 to 2018.
- ❖ Asia sold the highest dollar worth of exported onions during 2018 while exporters in Europe export onion at 34%, while 20% of worldwide shipments of onions appears from North America.
- ❖ Market sale/demand of onion from Africa is around 5.6%, Latin America 3.3%, Oceania 2.3%.

## 1.3 Identify bulb crop varieties as per day length requirement

### 1.3.1 Bulb Selection as per day length

- For growing good bulb onions, bulb crop cultivator needs to know about how day length and temperature affects onion plant growth.
- Day length or day light hours can be determined by sun tracking north in northern hemisphere during spring and summer while south during fall and winter which stimulates onion plant to start making bulb.
- Every onion variety forms a bulb after it has received certain number of hours of day light each day for certain number of days.
- Onions can be categorized into long-day, intermediate-day and short-day varieties.

### 1.3.2 Onion Bulb Formation in Day Light Hours

- Onion bulb development depends on day light as onion forms leafy tops and begins to form bulbs when each day light reaches at certain length.
- The amount of day light required for onion plant to forms a bulb varies with variety.
- The length of day light across world varies throughout the year as per latitude where it gets tilt of Earth's axis of rotation which makes year-long ecliptic journey around the sun.
- With solstice occurring about in June, north pole gets tilted towards the sun, so northern hemisphere has days ranging in duration from over 12 hours in southern part of northern hemisphere which is close to equator to about 24 hours at North Pole.
- In every hemisphere, high latitude lead to longer number of day light hours during summer and day light gets shorter during winter as location/latitude shows day length, where certain onion varieties are not suited for certain locations.

- Short day onions needs 12-14 hours of day light each day to form bulbs which happens in southern regions of northern hemisphere having latitude of not more than 36°N.
- Intermediate day onions needs 13-15 hours of day light where each day form bulbs that grows best between latitudes 35°N-38°N as many are adapted for production to latitude 42°N.
- Long day onion needs 14-16 hours of day light each day to form bulbs which are best in north of $36^{th}$ parallel or latitude 36°N.

## 1.4 Select appropriate bulb crop variety with maximum yield

### 1.4.1 Yield components

Yield parameters such as:

- harvest index
- total bulb yield
- marketable bulb yield
- unmarketable bulb yield
- total bulb number
- marketable bulb number
- unmarketable bulb number
- bulb quality parameters

Marketable Bulb Number Per Dunam:

It was obtained as mean number of marketable bulb numbers by counting good quality bulbs, better colour bulbs, medium bulbs, medium large bulbs and large bulbs per plot at harvest which was converted to dunam.

Unmarketable Bulb Number Per Dunam:

It was obtained as mean number of unmarketable bulb numbers by counting very small bulbs, under size bulbs, diseased bulbs over size bulbs per plot at harvest which was converted to dunam.

Total Bulb Number Per Dunam:

It was obtained as the sum of marketable bulb numbers per plot and unmarketable bulb numbers per plot at harvest which was converted to dunam.

Marketable Bulb Yield Per Dunam (Ton/Dunam):

It was obtained as mean weight of marketable bulb yield per plot at harvest by weighing bulbs in kg/plot which was converted to ton/dunam.

Unmarketable Bulb Yield Per Dunam (Ton/Dunam):

It was obtained as mean weight of unmarketable bulb yield per plot at harvest by weighing bulbs in kg/plot which was converted to ton/ dunam.

Total Bulb Yield Per Hectare (T/Ha):

It was obtained as sum of Marketable bulb yield per dunam (ton/dunam) and Unmarketable bulb yield per dunam (ton/dunam).

Harvest Index (%):

It is proportion of economic (bulb) yield to that of total biological yield which is the ratio of mature dry bulb yield of plants harvested from 3 central rows and total biological yield of plants of the same rows.

### 1.4.2 Bulb Quality Parameters

- Small Size Bulb Number: The small size bulbs number was obtained per plot at harvest by counting the number of bulbs/plot which was converted to ton/dunam.
- Medium Size Bulb Number: The medium size bulbs number was obtained per plot at harvest by counting number of bulbs/plot which was converted to ton/dunam.
- Large Size Bulb Number: The large size bulbs number was obtained per plot at harvest by counting number of bulbs/plot which was converted to ton/dunam.

## 1.5 Select bulb crop variety which is resistant to insect, pest and disease

### 1.5.1 Introduction

- Insect pests are responsible for substantial crop losses by direct damage and transmission of plant diseases.
- There are many strategies that improves crop resistance to insect pests which are related to plant secondary metabolism and immunity and microbiome science.
- Crop resistance to insects is cultural control methods which involves use of agronomic practices to reduce insect pest abundance and in damaging which occurs if there involves no practice.
- In integrated pest management, plant resistance to insects uses resistant crop varieties to suppress insect pest damage where plant resistance is used in conjunction with other direct control tactics.
- Crop resistance results as consequence of heritable plant qualities which result in plant that is being relatively less damaged as compared to plant without the qualities.
- The insect resistant crop cultivar yields more than susceptible cultivar when confronted with insect pest invasion.
- Resistance of plants is relative and is based on comparison with plants lacking the resistance characters such as susceptible plants.
- Insect resistant crop varieties suppress insect pest abundance or elevate the damage tolerance level of crop.
- The insect resistant plants alter the relationship which an insect pest has with its plant host.
- The use of insect resistant crop varieties is economically, ecologically and environmentally advantageous where economic benefits takes place due to crop yields that saves from loss to insect pests and money by not applying insecticides that would have been applied to susceptible varieties.
- The seed of insect resistant cultivars costs no more, or little more than for susceptible cultivars as ecological and environmental benefits arises from increases in species diversity in agroecosystem due to reduced use of insecticides.
- Increases in species diversity increases ecosystem stability which promotes more sustainable system far less polluted and detrimental to natural resources.

- ❖ The main advantage of using insect resistant crop varieties as a component of pest management arises from ecological compatibility and compatibility with other direct control tactics.
- ❖ Plant resistance to insect pests has advantages over other direct control tactics as plant resistance to insects is compatible with insecticide use, while biological control is not.

### 1.5.2 Onion Crop Disease Resistant

- ❖ Onion crop is attached by many diseases which causes heavy losses both in field and storage houses.
- ❖ Breeding for disease resistance remains a major activity at different centres where success achieved relates to identification of resistance sources for many diseases.

There are many onion crop related diseases in India:

Purple blotch

- ❖ The disease is due to Alternaria porri L. which affects the bulb and seed crop and is more serious in Kharif season.
- ❖ It occurs on leaves and flower stalks as small sunken whitish flecks with purple colours making the spots to enlarge and form large purple areas.

Basal rot

- ❖ It is caused by Fusarium oxysproum sp. cepae.
- ❖ Resistant genotypes to basal rot identified by various workers.
- ❖ Variations was reported towards susceptibility at seedling stage and later bulb development stage.
- ❖ Some of first showing resistance at seeding stage was found susceptible later at bulb development stage.

Stemphylium blight (Stemphylum vesicarium wall roth)

- ❖ In this, screening is done to locate resistance against the disease in terms of tolerant, combined tolerance to stemphylium leaf blight and purple blotch in 16 genotypes.
- ❖ It needs urgent in locating good source of resistance against such disease for developing resistant varieties.
- ❖ The other diseases which are of minor importance are downy mildew which is caused by Peronospora destructor and onion smut by Urocystis cepulae frost.
- ❖ Downy mildew was reported from hilly, besides plains on both seed crop and bulb crop.
- ❖ The losses caused due to such disease results in about 60-70%.

## 1.6 Select bulb crop variety which is draught tolerant

### 1.6.1 Introduction

Drought tolerant or drought resistant bulb crops are those that can withstand prolonged periods of dry weather and often have succulent leaves or stems and can sometimes store water. Other crops have fine hairs or silvery colouring of foliage that cuts down evaporation.

There are many suitable crops/plants that will cope in full sun and look good in a gravel setting having drought tolerant characteristics as silver or grey-green leaves to reflect sun and coating of fine hairs on leaves or stems, for trapping in moisture.

Bulbs are an excellent choice in the garden of the future, and with their natural moisture storing structure, most bulbs tend to be more drought resistant than plants without this adaptation.

And while many consumers are cancelling or postponing their landscaping plans over worries of water shortages and drought-raised irrigation costs, savvy homeowners and landscapers see the drought as an opportunity to replace the often mostly non-native, water-thirsty plants with more efficient gardens, using bulbs and succulents.

Bulbs are an excellent choice in the garden of the future, and with their natural moisture storing structure, most bulbs tend to be more drought resistant than plants without this adaptation. Many of these types of bulbs tolerate drought conditions, especially in the summer while dormant. Storing their entire life cycle in a swollen underground stem, the bulb contains the leaves, flowers, food and the bulblet.

While there are a number of drought tolerant vegetables varieties available, without some planning, extreme drought and heat will kill even the hardiest. Planting at the correct time is crucial. Sow the seeds earlier in the spring to take advantage of the warm weather and jump start the growing season, or plant later in the fall to minimize the use of irrigation and use the seasonal rains to your advantage.

### 1.6.2 Garlic and perennial onions

- Garlic and perennial onions are alliums with similar growing and storage requirements. Under the right conditions these are very productive crops, with harvests 3–8 times greater by weight than the amount planted. With the right care these alliums are self-sustaining, so they are particularly appropriate if you are interested in local food, permaculture and perennial crops.
- It is essential that the soil be kept moist during the entire period of active. If the soil is allowed to dry out just once, growth will be arrested, resulting in smaller bulbs and reduced yields. Heavy mulch between rows and plants will conserve moisture and control weeds.
- When planted at the proper depth, bulb clusters of multiplier onions form near the soil surface, not beneath the soil. This allows the bulbs to become larger, reduces susceptibility to rot, and improves keeping quality. Bulbs of garlic and topset onions develop from cloves beneath the soil surface.
- The onion plant was shown to be tolerant to water stress, being able to grow and produce seed yields in arid zones with limited amounts of rainfall and without irrigation. Under these conditions, large mother bulbs and high plant density gave the best results. Nitrogen was very important for the production of seed yield, but high amounts applied before planting and additional supply during the growing-season as top dressing did not have a significant effect on seed yield. Similarly, enrichment of the soil with phosphorus or potassium did not increase yield.

## 1.7 Select right quality and quantity of seed

### 1.7.1 Seed and its type

- The seed in a plant is the part that develops from the ovules after fertilization. They are enclosed in the fruit which develops from the fertilized ovary. The seeds are formed as a result of sexual reproduction and contain the young embryo which can develop into a new plant.
- A seed is a basic part of any plant. The ovules after fertilization, develop into seeds. A seed is made up of a seed coat and an embryo. The embryo is made up of a radicle, an embryonal axis and one (wheat, maize) or two cotyledons (gram and pea). A seed is found inside a fruit which converts into a new plant when we plant it. Hence, the seed is the most important part.
- There are hundreds of variations in the seed size, shape, colour and surface. The seeds range in size from tiny dust particles, as found in some orchids, to large double-coconuts. The seed surface may be smooth, wrinkled, striate, ribbed, furrowed, reticulate, tuberculate, alveolate, hairy, and pulpy or having patterns like finger prints.

Seeds can be recognized as:

- Nuclear seed
- Breeder seed
- Foundation seed
- Registered seed
- Certified seed

**Breeder seed**

- It is the seed or vegetatively propagated material which is directly controlled by originating or sponsoring breeder or institution serving as basic seed for recurring increase of foundation seed.

**Foundation seed**

- It is the progeny of breeder seed.
- The seed stock handled to maintain specific identity and genetic purity, which may be designated or distributed and produced under careful supervision of an agricultural experiment station.
- The seed is source of all other certified seed classes either directly or through registered seed.

**Registered seed**

- The progeny of the foundation seed maintain the genetic identity, purity, approval and certification by certifying agency.
- It serves as a quality which is suitable to produce certified seed.

**Certified seed**

- It is the progeny of the foundation seed.
- The production is handled to maintain genetical identity and physical purity as per standards as specified for crop certification.
- It should have minimum genetical purity of 99%.
- Certified seed have progeny of certified seed , provided the reproduction does not exceed two generations beyond foundation seed and provided if certification agency shows genetic and physical purity, if not be significantly altered.
- In highly self pollinated crops certification of further generation is permitted.
- Certified seed produced from certified seed is eligible for further seed increase under certification, except in case of highly self-pollinated crops, where certification of one further generation is allowed.
- Certification tags issued once for certified seed not eligible for further seed increase under certification.
- For paddy and wheat , certified seed produced from certified seed is eligible for certification by NSC up to two generations from foundation seed - Certified seed (I) - Certified seed (II)
- For barley, garden pea ,ground nut, soyabean, certified seed produced from certified seed is eligible for certification up to 3 generations from foundation seed - Certified seed (I) - Certified seed (II) - Certified seed (III)

### 1.7.2 Quality Seeds

**Introduction**

Seed is a material which is used for planting/regeneration purpose. It is a fertilized matured ovule which is covered with seed coat known as seed or propagating material which is part of agriculture, sericulture, silviculture and horticultural plants used for sowing or planting.

It is also defined as structurally true seed with fertilized matured ovule having embryonic plant, storage of food and protective seed coat, which has cotyledons and endosperm

In terms of seed technology, seed may be sexually produced matured ovule having intact embryo, endosperm and or cotyledon with protective covering. It is a propagating material of healthy seedlings, tuber, bulbs, rhizome, roots, cuttings, setts, slips and all types of grafts and vegetatively propagating materials used for production purpose.

It is most vital and crucial input for crop production, as it increase the productivity without adding appreciably to the extent of land under cultivation by planting quality seed.

Good quality seed is essential for production of crop suitable for export. Many farmers produce their own seed, which has been produced by Seed Certification Scheme or use Quality Declared Seed system. Such seed system has been designed to ensure the best possible quality seed production with the resources that are already available in seed production organizations.

The system is designed to provide quality control during seed production which is less demanding on government resources than fully developed seed certification, but adequate to provide good quality seed both within countries and in international trade.

**Characteristics**

Quality Seed must be genetically pure having:

- Breeder/nucleus: 100 %
- Foundation seeds: 99.5%
- Certified seeds varieties: 98%
- Certified hybrid seeds:95%
- Certified hybrid cotton: 90%
- Certified hybrid castor: 85%

Quality Seed should have required level of physical purity such as:

- For all crops: 98%
- Carrot: 95%
- Ragi: 97%

Quality Seed should be free from other crop seeds such as designated inseparable crop seeds that are plants seeds cultivated crops found in seed fields where seeds are similar to crop seed that is difficult to separate by mechanical means, causing physical admixtures with crop seeds only when these crop mature at same time when seed crop matures. It covers barley: oats and wheat seeds.

Quality Seed should be free from objectionable weed seeds, as seeds of weed species that are harmful in one or more of following ways;

- size and shape of weed seeds are similar to crop seed and difficult to separate them by mechanical means
- weeds growth habit is determinate type and compete with crop for all resources
- weed plant parts are poisonous or injurious to human and animals
- weed plants serves as alternative hosts for pests and diseases
- Ex. Bersem: chicory
- Paddy: wild paddy (Oryza sativa var. fatua)
- Cucurbits: Wild cucurbits spp.
- Lettuce: wild lettuce
- hendi: wild Abelmoscous spp
- Wheat: Convolvulus arvensis (Hirankuri)

Quality Seed should be free from designated diseases, such as diseases that are specified for seed certification causing contamination of seed lot

- Ex; loose smut of wheat (Ustilago tritici)
- Kernal smut of sorghum (Sphacelotheca sorghi)

- ❖ Grain smut of pearl millet (Tolyposporium penicillariae)
- ❖ Ergot of bajra (Claviceps microcephala)
- ❖ Early blight of tomato (Alternaria solani)
- ❖ Little leaf of brinjal (Datura virus-2) etc.,

### Attributes of seed quality

Seed quality is perhaps the most important aspect of seeds production, acquisition and use. A bad seed or seed lot is not only a waste of money, it is a waste of time and it brings mistrust among farmers, seed producers and seed sellers. Bad quality seeds affect the entire crop production value chain. Even more important is that bad quality seed with low germination power or count can discourage farmers from purchasing them especially where efforts are made for introducing improved seed varieties. There are four basic attributes of seed quality to consider:

1. Physical qualities of the seed in a specific seed lot
2. Physiological qualities observes germination and vigour of seeds
3. Genetic quality which relates to specific genetic characteristics of seed variety
4. Seed Health which refers to the presence of diseases and pests within a seed lot.

### Physical Quality

- ❖ It is the cleanliness of seed from other seeds, debris, inert matter, diseased seed and insect damaged seed.
- ❖ The seed with physical quality should be of uniform size, weight, and colour.
- ❖ The seed should be free from stones, debris, and dust, leafs, twigs, stems, flowers, fruit well without other crop seeds and inert material.
- ❖ Seed should be devoid of shriveled, diseased mottled, moulded, discoloured, damaged and empty seeds.
- ❖ The seed should be easily identifiable as species of specific category of specific species.
- ❖ Lack of quality character in seed will indirectly influence the field establishment and planting value of seed.
- ❖ The quality character of seed could be obtained with seed lots by proper cleaning and grading of seed after collection and before sowing / storage.

### Genetic purity

- ❖ It is true to type nature of seed.
- ❖ The seedling / plant / tree from the seed should resemble its mother in all aspects.
- ❖ The quality character of seed is important for achieving desired goal of raising the crop either yield or for resistance or for desired quality factors.

### Physiological Quality

- ❖ It is the actual expression of seed in further generation / multiplication.
- ❖ Physiological quality characters of seed have seed germination and seed vigour.

- The liveliness of seed is known as viability.
- The extent of liveliness for production of good seedling or ability of seed for seedling with normal root and shoot under favorable condition is known as germinability.
- Seed vigour is energy or stamina of seed in producing elite seedling which is the sum of all seed attributes that enables its regeneration under any given conditions.
- Seed vigour shows level of performance of seed or seed lot at germination and seedling emergence.
- Seed that do well at sowing are termed as quality seed and based on degree of performance in production of elite seedling, it is categorised as high, medium and low vigour seed.
- The difference in seed vigour is differential manifestation of deteriorative process which occurs in seed before ultimate loss of ability to germinate.
- Difference in seed vigour is expressed in rate of emergence, uniformity of emergence and loss of seed germination.
- The viable seeds need no germination but all germinable seed will be viable.
- The vigourous seeds will be germinable but all germinable seed need not be vigourous. Physiological quality of seed is achieved by proper selection of seeds used for sowing and by caring for quality characters during extraction, drying and storage.
- Seed with good vigour is preferable for raising a good plantation as the fruits, the economic come out are to be realized after several years.
- The selection of seed based on seed vigour is important for raising perfect finalize plantation.

**Seed Health**

- The health status of seed is the absence of insect infestation and fungal infection, in or on the seed.
- Seed should not be infected with fungi or infested with insect pests as these will reduce the physiological quality of seed and also the physical quality of the seed in long term storage.
- The health status of seed includes deterioration status of seed expressed by low vigour status of seed.
- The health status of seed influences the seed quality characters directly and warrants their soundness in seed for production of elite seedlings at nursery / field.

### 1.7.3 Seed Selection

- The quality seed should be used for getting healthy seedlings. The crop yield to tally depends on the seed quality. In seed crop, grower should aim for higher yield without sacrificing genetic purity which is more important. The yield and quality of seeds depends on seed selection and its variety.

Selection of variety:

- Variety should be adopted to agro-climatic condition

- ❖ Variety should be high yielder
- ❖ Variety should posses other desirable characters like disease resistance, earliness, grain quality etc

Selection of Seed:

- ❖ Seed purity: The seed used for raising a seed crop should be of known purity, appropriate class and invariably obtained from authorized official agency.
- ❖ Seed Treatment: If the seed is not treated already, it should be treated with appropriate fungicides/insecticide prior to sowing.
- ❖ Time of sowing: Should be sown at their normal planting time
- ❖ Seed rate: Lower seed rate than usual to facilitate rogunig operation/seed inspection.
- ❖ Method/depth of sowing: Invariably sown in rows at right depth.

## 1.8 Identify various certified seed vendors/suppliers

### 1.8.1 Seed producing organizations

There are three types of Seed Producing Agencies such as:

- ❖ Government sector
- ❖ Public sector
- ❖ Private sector

**Public sector organizations include:**

**National Seeds Corporation (NSC)**

NSC was established in 1963 which involves in work of quality seed production. Its responsibilities are defined with the inception of National Seeds Project (NSP) and its main function involves in:

- ❖ Production of Breeder Seed: The requirement of breeder seed of various crops is assessed and arrangements are made for its production through crop research institutes of ICAR and State Agricultural Universities.
- ❖ Production of Foundation Seed: The production and distribution of the foundation seed of the varieties of national importance is the responsibility of the NSC.
- ❖ Production of Certified Seed: The production and marketing of the certified seeds of vegetables, oilseeds, pulses and fodder crops is taken up.
- ❖ Marketing: Interstate marketing of the seeds of national varieties lies with the NSC.
- ❖ Consultancy Services: The NSC provides consultancy services related to designing, procurement, and installation of seed processing plants, equipments, etc.
- ❖ Imparts Training: The NSC organizes short-term training courses to the personnel engaged in seed production work.
- ❖ Coordination: NSC provides coordination in the production of certified seed by SSCs.

- Taking up seed certification work in those states where SSCA has not been established.
- Import and export work of seed.

**State Seed Corporations (SSCs)**

National Seed Project in 1976 establishes State Seed Corporations in 14 states of India. Its functions involve in production and distribution of certified seed in respective state. With the establishment of Tarai Development Corporation in U.P. in 1969 for production of certified seeds, SSCs has been established in other states based on similar pattern to faster the development of seed industry in the country. NSP provides finance for establishment of new SSCs and strengthening of the existing ones. The NSP provides funds for following purposes:

- Strengthening seed production, processing and marketing facilities of NSC.
- Strengthening the facilities of the NSC for producing and processing seeds of vegetables, fodder crops and other specialised crops not taken up by the SSCs.
- Strengthening the facilities for nucleus, breeder and foundation seed production; processing, and storage at selected Central Research Institute of ICAR and at SAUs.

State Seed Certification Agencies (SSCAs): State Seed Certification Agencies are responsible for seed certification in states. SSCAs performs following functions:

- screen applications from seed growers for seed certification and decides on their fitness,
- check and verify appropriateness of source seed used for growing seed crop under certification,
- do requisite field inspections,
- conduct seed tests,
- certify seeds found suitable and issue appropriate tags for certified and foundation seeds,
- guide seed growers on production, processing and distribution of seeds,
- conduct short courses on seed production for seed growers,
- participate in activities conducted to development of seed industry,

**Private sector includes various seed companies.**

There are many private seed companies which involves in work of quality seed production such as:

- Maharashtra Hybrid Corporation, Jalna, 1964.
- Pioneer Seed Company, Hyderabad.
- Nath Seed Company Pvt. Ltd., Aurangabad.
- Hindustan Lever Ltd., Mumbai.
- Proagro Seed Company, New Delhi.
- Ankur Seed Company, Nagpur.

- ❖ Vijay Seed Company, Jalna.
- ❖ Mahendra Seed Company, Jalna.

## 1.9 Procure seed and prepare the nursery

### 1.9.1 Seed Procurement

Emergency interventions involving seed should be based on assessment regarding the overall livelihood status of affected populations. If assessment shows a problem related to seed security then, appropriate strategy to address the problem can be implemented. Seed relief can be significant in many cases as it helps vulnerable population in acute emergencies to move from receiving food aid to producing food which improves food security and reducing vulnerability. In chronic emergencies due to drought, benefit of providing seed when there is high probability of renewed crop failure is big task.

There are many ways for obtaining seed:

- ❖ local procurement
- ❖ market-based approaches
    - For traditional varieties/local landraces of field crops, local procurement or market based approaches gives preferred option to ensure that right crops and varieties are purchased and provided to farmers.
    - International procurement are normally used when seed is not locally available.
    - Increasingly local procurement and market-based approaches are used for emergency operations and new methods like seed vouchers are being developed.
    - In the rehabilitation phase, seed multiplication is done at community level to build more sustainable seed security.
    - Procurement in country works with national seed systems whereas international procurement can undermine national seed systems.

Local procurement

- ❖ Local procurement which is under the authority or designated official follows a tender process through which, seeds can be procured.
- ❖ Local procurement is necessary when dealing with field crops where issue of varietal suitability becomes extremely important.
- ❖ In many areas, farmers normally plant local landraces that they prefer and are well adapted to local conditions.
- ❖ The traditional varieties or landraces are referred as populations of plants as their heterogeneous characteristics, rather than pure homogenous crop varieties.
- ❖ Landraces are difficult to purchase outside the country or even outside particular agro-ecology of country as varieties of field crops, international tendering may not yield good response since the local varieties are often not known outside the borders of the importing country.
- ❖ In local procurement, landraces are preferred, but there is no clear distinction between seed and food grain in markets which leads to food grain being used for seed security activities.

- ❖ Seed for planting must be able to germinate and produce a good crop, so some seed suppliers clean, sort or process food grain and sell it as seed to relief organizations.
- ❖ The physical aspects of seed quality can be discerned by observation but genetic and physiological aspects must be determined by complex tests that can not be performed quickly.

International procurement

- ❖ The procedures for purchasing of seed valued at more price is done by the Procurement and purchasing Branch which involves in detailed specifications for seed and packaging materials with delivery instructions.
- ❖ The bidder should satisfies the technical specifications of the tender with competitive price and proposes an acceptable delivery time.
- ❖ After the selection of bidder, seed technical specifications and lowest price, the seed will be sampled by sampling company and tested at accredited laboratory for critical seed attributes specified by seed technical unit before shipping.

### 1.9.2 Nursery

Nursery is a place where plants are grown, take care of and sold out. It is also a place where plants, shrubs and ornamental trees are grown for transplanting into landscape areas.

**Fig Plants grown in nursery in pot/container**

- ❖ In Nursery, plants can be grown on fields or in a pot/container. Plants that are grown in pot/container are of shorter to medium height and can be easily movable.

### Categories

Nurseries can be categorized as:

1) Temporary nursery

- ❖ Such type of nursery is developed only to fulfill the need of season or targeted project.
- ❖ Here, production of seedlings of transplanted vegetables and flower crops are temporary in nature.
- ❖ In this, temporary arrangement of growing of forest seedlings for planting in an area is done.

Advantages:

- ❖ It is can be shifted to any site.
- ❖ It needs low capital cost
- ❖ It reduces transportation cost

- ❖ It works within climatic site conditions
- ❖ In this, skilled labour is not required
- ❖ It can be near the planting site

Disadvantages:

- ❖ In this, water supply can be poor
- ❖ Soil is also poor in this
- ❖ There appears shortage of trained labour
- ❖ All can result in poor quality trees

2) Permanent nursery

- ❖ Such nursery is placed permanently so as to produce plants regularly.
- ❖ The nursery has all permanent features and has permanent mother plants.
- ❖ In such type of nursery, the work can be done on regular basis all the year round.
- ❖ With the type of plants produced, nurseries can be classified as:
    - Fruit plant Nurseries: In this, seedlings and grafting of fruit crops are prepared.
    - Vegetable nurseries: In this, seedlings of cauliflower, cabbage, brinjal and tomato are prepared.
    - Flowers plants nurseries: In this, seedlings of flowering plants like gerbera, carnation, petunia, salvia, rose, chrysanthemum, coleus, aster, dianthus are developed.
    - Forest nurseries: In this, seedlings of plants which is useful for forestation like pine, oak, teak, eucalyptus, casuarinas are prepared and sold.
    - Different nurseries: In this, plants having high economic value, rare and medicinal, herbal plants are propagated such as geranium, rose, calendula and marigold.

Advantages:

- ❖ It is well designed
- ❖ It has a good water supply
- ❖ It provides good quality and required number and size of plants
- ❖ It has good soil
- ❖ It can be well managed and protected
- ❖ In this, planting material is always available

Disadvantages:

- It requires high capital and transportation cost
- It is not always in climatic site conditions
- It is not always near the planting site
- In this, permanent and skilled labour is required.

3) Dry Nursery

- It is a type of nursery that is maintained without any irrigation or other artificial watering.

4) Wet Nursery

- It is a type of nursery that is maintained by irrigation or other artificial watering during the dry periods.

**Importance**

A nursery is a part of agriculture where plants are propagated, nurtured, grown, and sold out for the home garden or commercial purpose. The main purpose of nurseries is to supply young plants or saplings to gardens, farms, agriculture, forestry, and for conservation. Nurseries can have stock of various plants such as herbs, shrubs, plants, small trees, big trees, and bonsai. A nursery plant center offers to its customer such as home gardener, landscape gardener, or commercial farmer, all types of young plants that are native and imported.

Establishment of nursery is important for production of:

- proposed size and age of seedlings.
- desired number of seedlings.
- healthy plants of the desired spp.
- plants of superior quality.
- desired plants at right time.
- seedlings at a cheaper cost.

Importance

- In nursery, seedlings and grafts are produced from which the fruit orchards and ornamental gardens can be established with minimum care, cost and maintenance.
- The nursery planting materials are available at the star of planting season which helps in saving time, money and efforts of farmers to raise seedlings.
- There is a wide scope for fruit orchards, ornamental, vegetable, and landscape gardens at public places, highways and co-operative housing societies.
- It assures the production of genetically improved quality planting material.
- It provides employment opportunities for technical, skilled, semi-skilled, unskilled labour.
- They are important source for supply of seedlings for fruit, pulp and paper, fuel wood, timber and other industrial demands.

## Nursery site

The selection of the area for a nursery is critical and it should be sited as centrally as possible to the field(s) to be planted. Nurseries can be started irrespective of where you are located but with enough land space to grow plants. Nursery location may be predetermined based on cost investment, availability of property. The choice of Nursery location will directly influence on marketing the plant saplings, manpower and future expansion. Location of nursery must be near to major cities and townships, transportation, availability of labor and convenient for future expansion. Nursery site location must be nutrient-rich or medium soil with good water source. Nursery site must have good soil drainage and free from soil pathogens.

There are certain factors which need to be considered while selecting a nursery site such as:

Topography (Terrain)

- ❖ The selected area should be flat with gently slopes between 0 and $3^0$
- ❖ The area should have reliable/permanent source of water supply for irrigation

Water Supply

- ❖ Water requirements like quality and quantity should be determined before starting of site preparation.
- ❖ The easiest source of water is where there is a large natural pond or lake whereby all that is required is to place an intake pipe to connect with a pump unit.
- ❖ A back-up system should be considered, mainly in isolated areas or areas of lower or unreliable rainfall.

Drainage

- ❖ The site selected should not be prone to flooding, as it will damage the seedlings and buildings/stores.

Area/Spacing

- ❖ To have good growing conditions with minimal risk of etiolation, nursery should have a planting density of 13,800 polybags per ha with 0.91m(3ft) x 0.91m(3ft).
- ❖ The spacing should be increased by 0.15m (0.5ft) if seedlings are anticipated to be kept in the nursery for more than 12 months.

Accessibility/Approach

- ❖ Walk passage in nursery and their alignments should be carefully planned and laid as per the placement distance of polybags and type of irrigation to be used.
- ❖ Approach to nursery should be sufficiently wide to allow vehicles to move during peak planting periods to facilitate supervision and movement of materials.

## Growing Media

- ❖ A growing medium serves four functions:

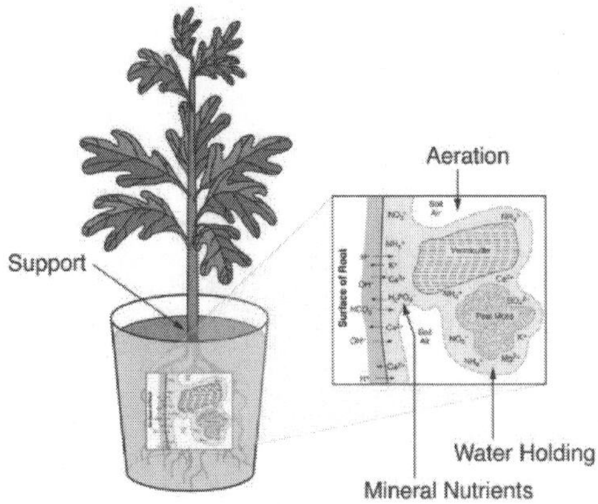


1. Physical Support

- ❖ The growing medium must be porous but should provide physical support.
- ❖ Young plants are fragile which should remain upright so that they can photosynthesize and grow properly.
- ❖ With larger nursery items in single containers, growing medium must be heavy enough to hold the plant upright against the wind.

2. Aeration

- ❖ Plant roots need a steady supply of oxygen to convert the photosynthate from the leaves into energy so that the roots can grow and take up water and mineral nutrients.
- ❖ The byproduct of such respiration is carbon dioxide which should be dispersed into atmosphere to prevent the buildup of toxic concentrations in root zone.
- ❖ The gas exchange from this takes place in large pores or air spaces in growing medium as nursery plants grow rapidly, which needs a medium with good porosity.

3. Water Supply

- ❖ Nursery plants use large amount of water for growth and development, so water supply should be provided by growing medium.
- ❖ Growing media are formulated so that they can hold water in small pores between their particles.
- ❖ Many growing media has high percentage of organic matter like peat moss and compost as these materials have internal spaces that can hold water like sponge.
- ❖ The growing media should have adequate porosity to absorb and store large amounts of water needed by growing plant.

4. Supply of Mineral Nutrients

- Many essential mineral nutrients that nursery plants needed for rapid growth should be obtained by roots from growing medium.
- Many mineral nutrients are electrically charged ions while positively charged ions (cations) include ammonium nitrogen ($NH4^+$), potassium ($K^+$), calcium ($Ca^+2$), and magnesium ($Mg^+2$).
- The cations are attracted to negatively charged sites on growing medium particles up to the point when roots extracted cations.

## Propagation

Plant propagation is science and an art where science of plant propagation deals with know how of plant physiology, nursery cultural practices, and characteristics of particular plant that a gardener needs to grow. The art of plant propagation cannot be taught in book or classroom, as it consists of specific technical skills that must be achieved by innate ability or experience and often needs certain "feel." Good plant pagators are said to have a "green thumb."

Plants are propagated either by seed or by vegetative means, such as by cuttings, grafting or tissue culture. Some nurseries specialize in growing and selling prefinished plants to other growers, making propagation their sole business. However, some wholesale nursery operations have their own propagation areas where plants are produced for in-house use.

Plant propagation is the process of creating new plants. There are two types of propagation: sexual and asexual. Sexual reproduction is the union of the pollen and egg, drawing from the genes of two parents to create a new, third individual. Sexual propagation involves the floral parts of a plant. Asexual propagation involves taking a part of one parent plant and causing it to regenerate itself into a new plant. The resulting new plant is genetically identical its parent. Asexual propagation involves the vegetative parts of a plant: stems, roots, or leaves.

In nursery, native plants can be produced in either container nurseries or bareroot nurseries. Although some natives are produced bareroot, most are grown in containers as they offer more flexibility in scheduling, and can be grown in shorter period. In particular, size and shape of native plant seeds limits how they can be grown. Most commercial conifers have relatively small, smooth seeds that can be easily sown in bareroot seedbeds. Most other natives, however, have seeds that are very small, irregularly-shaped, or have appendages that make mechanized sowing difficult if not impossible. All these characteristics make container propagation much more attractive.

## Chapter 2

# Soil Preparation and Plantation of Bulb Crops

## 2.1 Identify soil types and it should be tested in state govt. approved laboratory

### 2.1.1. Soil Type

Soil is one of the important resources, as fertile soil helps in producing many crops which serves the food requirements not only in the country but also in other parts of the world. Soil is not same at every place on the earth. The types of soil of a place are determined by climate, landscape, and vegetation of that place.

Different types of soil are characteristics of their color, texture and chemical properties. The soil has particles of different sizes. The relative amounts of the size of various particles in the soil determine the texture of the soil.

#### 1. Sandy soil

- ❖ Sandy soil contains more than 60% sand and clay.
- ❖ It contains very little clay and silt, so it is porous.
- ❖ The size of soil particles in sandy soil is from 0.2mm to 2.0mm.
- ❖ The water building capacity of a sandy soil is very poor, so there appears lot of air in such type of soil.

- ❖ The soil is not good for plants.
- ❖ If water is available for irrigation then crops such as maize, millets, barley can be grown in desert soil.
- ❖ Cactus also grows in this soil.

**2. Clayey Soil**

- The clayey soil consists of very fine particles of clay.
- Its water holding capacity is very high.
- Wet clay soil is very sticky.
- It contains very little air.
- The size of soil particles in clay size is less than 0.2mm.
- The soil is rich in organic matter.

- The soil is not good for many plants.
- It is only good for crops like paddy, which require a lot of water.
- Clay is used for making toys, pots, and many other purposes.

**3. Loamy Soil**

- The loamy soil consists of sand, clay and silt.
- It contains enough hummus.
- It has a good water-holding capacity.
- It has sufficient aeration.
- It is well suited for cultivation.
- Roots of plants get enough water, air, and space to grow.

- The soil is ideal for growing crops such as wheat, sugarcane, cotton, jute, pulses, and oilseeds.
- Vegetables also grow well in this soil.

**4. Red soil**

- The red color of soil is due to the presence of iron oxide.
- It appears yellow when it contains less iron or more water.
- The soil contains a mixture of clay and sand, antis not fertile.
- It can be fertile by adding manures and fertilizers.

- It is suitable for growing groundnuts, pulses, millet, cotton and tobacco.

**5. Black soil**

- Black soil is also known as black lava soil.
- This soil black in color.
- It is formed from lava rocks and is rich in clay.

- It is ideal for growing crops such as cotton, sugarcane, tobacco, wheat, millets, and oilseeds.

## 2.1.2 Soil Composition

Soil contains:

- Organic matter
- Soil organisms - Micro flora and Micro fauna
- Soil water
- Soil air
- Inorganic matter - Macro nutrients and Micro nutrients

Organic Matter

- The plants and animals grown in weathered material and the organic residues left behind decay with time and become an integral part of the soil.
- The main source of soil organic matter is plant tissue.
- Animals are subsidiary source of soil organic matter.
- The micro flora like bacteria, fungi, algae, actinomycetes, and micro fauna like protozoa, nematodes, macro fauna etc, plays an important role in formation of organic matter.
- The organic matter influences the soil in respect to colour, physical properties, supply of available nutrients and adsorptive capacity.

Soil Organisms

- Soil is the habitat for enormous number of living organisms.

- Some of the organisms are visible to naked eye while others can be seen by microscope only.
- Roots of higher plants are considered as soil macro flora while bacteria, fungi, algae and actinomycetes are considered as soil micro flora.
- Protozoa and nematodes are the significant soil micro fauna where as the earthworms, moles and ants constitutes soil macro fauna.

Soil Water

- To function as a medium for plant growth, soil should have some water.
- Promotes many physical and biological activities of soil.
- Acts as a solvent and carrier of nutrients.
- Serves as a nutrient itself.
- Acts as an agent in photosynthesis process.
- Maintains turgidity of plants.
- Acts as an agent in weathering of rocks and minerals.

Soil Air

- Oxygen is required for all biological reactions that takes place in soil.
- The gaseous phase of soil acts as a path way for intake of oxygen which is absorbed by soil micro organisms, plant roots and for escaping carbondioxide by plants.
- The two way process is called soil aeration which become important for plant growth when water content is high, as water replaces soil air.

Soil Inorganic Matter

- The inorganic constituents of soil has carbonates, soluble salts, free oxides of iron, aluminium and silica with some amorphous silicates.
- The inorganic constituents forms bulk of solid phase of the soil.
- Soils with more than 20% of the organic constituents are designated as organic soils.
- Soils where inorganic constituents are called mineral soils.
- The majority of the soils in India are mineral soils.

## 2.1.3 Soil Testing

### Soil sampling

Soil sampling is done for checking of soil pH, phosphorus, potassium and micronutrients which should be done after every three to four years as these values do not change rapidly even with fertilization. Soil samples for soil nitrate should be taken 1-3 months before planting each wheat crop. In most of dryland or rainfed wheat production areas, wheat root will have depth of 5-foot with full profile of soil water.

Effective and profitable fertilizer recommendations require accurate sample of the field. It is very important that the soil sample should accurately represent the field to be fertilized. Previous crop history, previous lime and fertilizer rates and soil characteristics will cause variance in results if not accounted while sampling.

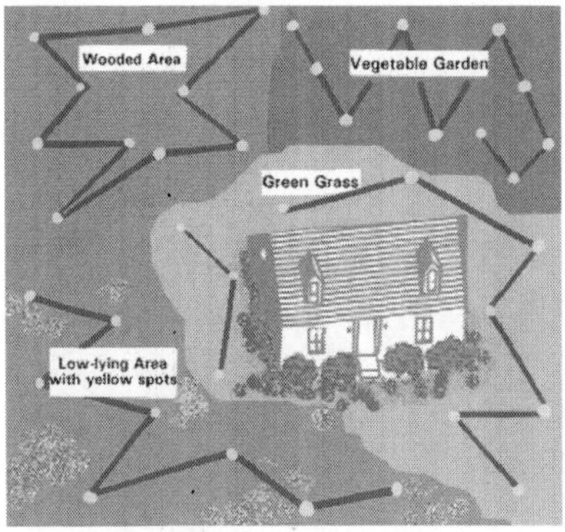


- ❖ Soil sampling is most challengeable work, as few grams of soil sample represents for a given area, hence soil samples needs to be taken in such a manner that the collected sample should reflect the true fertility of soil for any part of the land area.
- ❖ If the field is levelled and soil appears to be uniform, only one composite sample if taken properly could be enough for an area of 4-5 ha.
- ❖ Variation in slope, colour, texture, crop growth, unusual spots and management practice should be taken into account and separate sets of composite samples must be collected from each such area.
- ❖ Recently fertilized plots, bunds, channels, marshy tracts and area near trees, wells, cattle dung and compost piles or other non-representative locations must be avoided during the sampling.

### Preparation of soil samples

Preparing the collected soil sample for testing in laboratory involves the following steps:

- ❖ Soil samples which are collected from different units usually contain moist, aggregated state, etc that is not suitable for laboratory analysis.
- ❖ The soil sample is collected in jar/bucket should be free from organic debris, rocks and trash and needs to be mixed properly.

- ❖ Pour the soil sample from the bucket/jar on a piece of clean paper or cloth and mix it thoroughly.

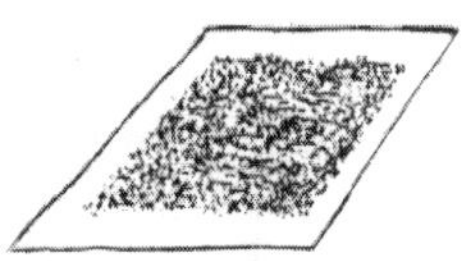

- ❖ The soil sample spread on a plastic or on thick paper should be dried and then grinded to avoid contamination.

- Field moist samples before drying, should pass through a 6 mm sieve by rubbing with fingers.
- After sieving, samples are air-dried (25-35° C; relative humidity 20-60%) and stored.
- Samples should not be exposed directly with the sunlight, but should be kept under the shade for drying.

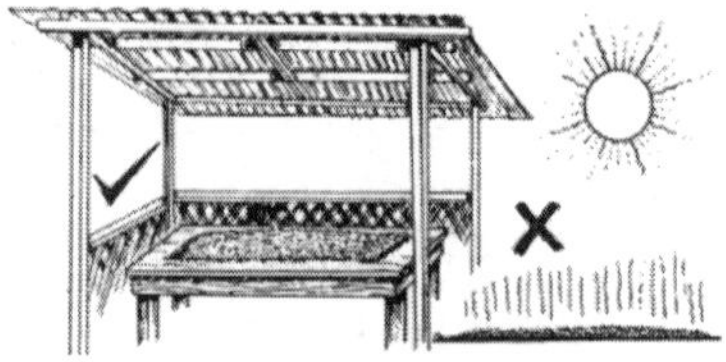

- To speed up the drying process, break the samples into different compartments.

- Keep the dried soil sample in a clean plastic bag free from any contamination.
- Write details of the sample in the information sheet and put it in the plastic bag.
- Tie the mouth of the bag carefully.

### 2.1.4 Soil Testing in Laboratory

Soil test refers to one or more of wide variety of soil analysis which is conducted for one of several possible reasons. Soil test shows analysis of soil sample which determines nutrient content, composition, and other characteristics like acidity or pH level. A soil test determines fertility, or expected growth potential of the soil which shows nutrient deficiencies, potential toxicities from excessive fertility and inhibitions from presence of non-essential trace minerals.

Soil testing is performed by commercial labs that offers variety of tests, targeting groups of compounds and minerals. It involves collecting of soil samples, preparation for analysis, chemical or physical analysis, interpretation of analysis results, and finally making fertilizer and lime recommendations for crops. The laboratory tests often checks for plant nutrients in three categories such as:

- Major nutrients: nitrogen (N), phosphorus (P), and potassium (K)
- Secondary nutrients: sulfur, calcium, magnesium
- Minor nutrients: iron, manganese, copper, zinc, boron, molybdenum, chlorine

Soil testing is an important diagnostic tool which determines nutrient needs of plants and for environmental assessments. It facilitates fertilizer composition and dosage selection for land employed in both agricultural and horticultural industries. Soil testing can also be used to identify application rates of waste materials containing nutrients or other elements that could harm the environment.

Soil testing is the best method available for determining how much nutrient is needed for optimum production of most crops. For certain crops, use of plant analysis serves as preferred method of determining crop nutrient needs. Making nutrient-application decisions without soil testing or plant analysis can be a risky endeavor because this will probably result in reduced profitability from either under or over-applying nutrients; in addition, over-application of nutrients can result in reduced water quality. It is important to follow recommended procedures for taking soil and plant samples.

## Facilities/Equipments in Laboratories

Soil sampling laboratories have following physical facilities:

a. Soil preparation room

b. Sample storage space

c. Main laboratory area

d. Research room

e. Chemical storage room

f. General office and hall

Soil sampling laboratories have following equipments and instruments:

1. Conductivity bridge
2. Shaking apparatus
3. PH meter
4. Rincing and stirring
5. Photoelectric colorimeter
6. Flame photometer
7. Centrifuge
8. Analytical balance.
9. Triple beam balance
10. Multiple dispensing equipment
11. Washing assembly
12. Pipetting assembly
13. Calculating machine
14. Constant holtage stabilizer
15. Water distillery
16. Oven
17. Air compressor

Materials and Equipments required in soil sampling laboratories include:

a. samples sheets with result.

b. Electrodes for PH meter.

c. Glass atomizer for flame photometer.

d. Conductivity Cell

e. Lamps – for colourimeter

f. Filter for colourimeter

g. Standard solutions for phosphorous, potassium and organic carbon.

h. Glasswares

i. Chemicals

j. Weighing scoop

k. Campe hairbrushes

l. PH meter

m. Colourimeter

n. Flame photometer

o. Conductivity bridge

p. Vacuum pump

q. Platining kit

r. Shaking machine

**Soil Test Report**

- Many laboratories involves in standard soil test which covers set of soil chemical measurements that are important in agronomic decision making.
- Certain labs vary, typically which includes pH, buffer pH, cation exchange capacity (CEC), base saturation, organic matter (OM), and the macronutrients phosphorus (P), potassium ($K^+$), calcium ($Ca^{2+}$), and magnesium ($Mg^{2+}$).
- In plains and western regions where salinity and sodicity are more common, conductivity, soluble salts, sodium ($Na^+$) and chloride ($Cl^-$) may also be reported in a local laboratory's standard test.
- Nitrate ($NO^{3-}$) and sulfur (S), are not usually part of standard tests, but are considered macronutrients.
- Nitrate tests are used in some regions for preplant nitrogen (N) management and in-season sidedress decisions, but recommendations vary from region to region.

## 2.2 Plough the main field for achieving recommended tilth

### 2.2.1 Tillage

- Tillage is an agricultural preparation of soil using mechanical agitation such as digging, stirring and overturning.
- The human-powered tilling methods with hand tools include shovelling, picking, mattock work, hoeing, and raking.
- Tillage has been used for millennia to prepare soil before sowing many of annual grain crops which involves applying power to break up and rearranging complete topsoil structure.
- The main aim of tillage is to destroy weeds and pests and also important for incorporating, redistributing or releasing nutrients and making soil texture suitable for seed sowing, seed germination and easy penetration of seedling roots.
- Tillage operation and cultivation in farming season comprises of plowing, preparing seed bed, leveling, cultivating, covering seed, making irrigation furrow, and application of fertilizer.
- Machines used for tillage operation consist of different kinds of plowshare, disk, harrow, cultivator, leveler, planter and furrow opener that gets attached to tractor and are used in farm to do various operations.

### 2.2.2 Need of till

- Soil disturbance before or during seed placement ensures that sowing depth is optimal and consistent.
- To ensure that seedling coleoptile can penetrate its covering of soil and roots rapidly to explore soil to its depth.
- Seeds are sown at wrong depth or at irregular depth due to poorly prepared cloddy soils or inadequate seeding equipment which gives poor yields.
- The aim of soil disturbance is to set back or kill any weeds and pathogens that will restrain growth of developing seedlings.

### 2.2.3 Effects of tillage

- It loosens and aerates top layer of soil/horizon, which facilitates planting the crop.
- It helps in mixing harvest residue, organic matter (humus) and nutrients evenly in the soil.
- It mechanically destroys weeds.
- It dries soil before seeding.
- It helps in exposing soil crumble over winter through frosting and defrosting, that prepares smooth surface for spring planting.

## 2.3 Requirement of seed per ha

Onion seeds are sown in nursery beds to raise seedlings. For onion:

- 8-10 kg of seed is sufficient to raise the seedlings for one hectare, which accommodates 100-110 beds of 3m X 0.6m size each.
- For broadcasting directly in field or sowing in rows, around 15-20 kg of seeds are enough for 1 hectare.
- For multiplier onions, around 10-12 quintal bulblets are required for planting one hectare.

Garlic crop cultivation is one of the most important bulb crop cultivations in India. For garlic:

- Plantation of 16 cloves per bed foot on 40 inch beds center-to-center is required.
- Around 400-500 kg of garlic seed per hectare is sufficient.
- Garlic is cultivated in Rabi and summer days which can be done by August-November.
- With 50-70 q per hectare, garlic per hectare is obtained.

## 2.4 Prepare nursery bed for seedling preparation

### 2.4.1 Introduction

Onion is a cool season vegetable crop which is cultivated on large scale chiefly during rabi season but also grown in some states during Kharif season.

### 2.4.2 Nursery Bed Preparation

- Onion seeds are sown on nursery beds to raise seedlings for transplanting in field.
- Raise the beds of size 3 x 0.6 m and 10-15 cm in height.
- Keep distance of around 70 cm between the two beds for carry out operations such as watering, weeding, etc.

- The surface of beds should be smooth and well levelled.

- Apply 15-20 gm of single super phosphate at the time of bed preparation for each bed.
- Raise the beds such that it avoids problem of water logging in heavy soils.
- The beds should be irrigated for 15-20 days, before sowing and should be covered with 250 gauge transparent polythene for soil solarisation.
- In sandy soils, sowing can be taken in flat beds.
- Seed bulbs should be treated with thiram at rate of 2 gm/kg of seed.
- Sowing of seeds should be done in lines at a proper spacing 5-7 cm row-to-row.
- To avoid mortality of seedlings due to damping off, drenching of beds with Bavistin is effective.
- Seeds after sowing should be covered with fine powdered compost followed by light watering.
- The beds should be covered with dry straw or grass or sugarcane leaves to maintain required temperature and moisture.

## 2.5 Proper size and number of nursery bed required for one hectare

### 2.5.1 Introduction

- Tillage is used to prepare land for planting.
- The only part of seedbed that needs clod-free is narrow row zone where seeds are planted.
- Seedbed preparation is location-specific and varies with climate, soil type, crop, management level and equipment.
- There are three types of seedbeds which include flat beds, raised beds and sunken beds.
- The width of raised seedbed is around 100-130 cm wide while the width of sunken seed is about 100-130 cm which is 2-5 cm below the surrounding soil level.
- The seedbed width ranges from 100-130 cm, which allow all plants and bed area to easily approach without walking in bed.
- The seedbeds should be kept to 10 meters or less in length so that alleyways can run up and down and across the garden to facilitate access.
- The seedbed alleyways should be narrow as 50-60 cm, but wider seedbed alleyways should be around 100 cm for wheelbarrow traffic.
- In seedbed, as there is no need to walk, so row spacing can be based on required plant spread.

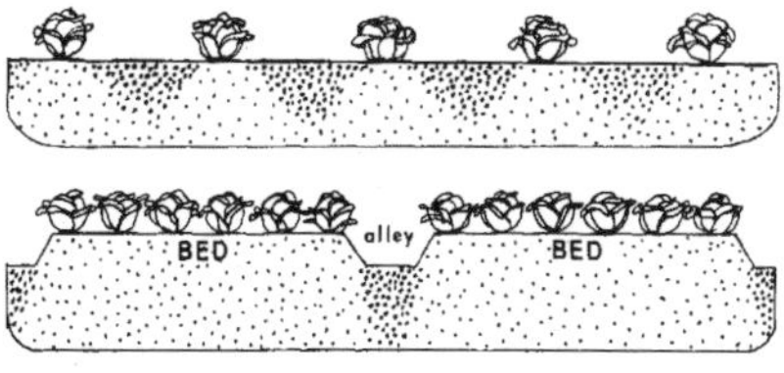

- ❖ For intensive layout of seedbed, 65 sq. meters of actual bed area and 35 sq. meters of alleyway area per 100 sq. meters should be there.
- ❖ For 1 metre square area, 5-7 gm of seeds would suffice and for planting in 1 hectare, 7-8 kg of seeds are required.

### 2.5.2 Nursery Bed for Garlic

- ❖ Garlic cultivation is done in moderate sandy loam soil.
- ❖ Soils with good quantity of potassium is considered good for garlic cultivation.
- ❖ Garlic crop is grown under the ground, whose roots penetrates to maximum of 20-25 cm from the ground.
- ❖ Soils that are rich in organic matter with a pH of 6.0-6.5 are best.
- ❖ Loamy soils with good drainage are considered good for garlic cultivation.
- ❖ For garlic, heavy soil is not good, since it does not have proper drainage as a result of which, tubers remain small.
- ❖ For garlic cultivation, initial plowing rotates upper 15-20 cm of soil so deep plowing should be done.
- ❖ Earlier than plowing, cultivator should be used twice or thrice, so that the lumps get broken and soil becomes frostbitten.
- ❖ In fields, add 20-25 tons of well-mixed dung compost and mix it well with harrow in soil.
- ❖ Planting garlic in well-prepared fields, types of land, weather, amount of rainfall etc is important.
- ❖ For preparation main field, he land should be ploughed to fine tilth to form ridges and furrows at 30 cm spacing or beds of convenient sizes.
- ❖ The cloves are spaced at 10 cm apart in rows with distance of around 20-30 cm as per size of bulb.

### 2.5.3 Nursery Bed for Onion

- ❖ Onion seeds require firm, finely textured soil in seed bed for good germination.
- ❖ Onion grows best in soil with good organic matter with pH of around 6.0-7.8.

- Seed-bed preparation of onion starts in fall of previous year where field is plowed to improve soil condition.
- After plowing, soil is shaped into beds that are settled by freezing and thawing action during winters.
- In spring, before planting, beds are smoothed with bed shaper harrow or roller and are planted.
- Due to certain onion diseases in soil, fields should be planted to onions only once every 5-7 years.
- Fields should be seeded from early March through mid-April when onion beds are dry enough to avoid compaction or germination problems during planting.

## 2.6 Seed and soil treatment before sowing

### 2.6.1 Pre sowing

Zerp till-drill

Bed planting

PTOS

Farmer's practice (FP)

A pre-sowing was applied for proper seed germination where:

- A high yielding variety of seed can be sown using zero till drill, bed former, PTOS and farmer's practice at the rate of 120 kg/ha.
- Seeding is done without land preparation by zero till-drill.
- Bed can be formed and seeding can be done without prior land preparation.
- Bed was formed two rows accommodating wheat.
- Furrow to furrow was 55-56 cm.
- Bed height was 12-13 cm.
- Irrigation water was applied to furrows between beds.
- Ploughing, seeding, and leveling were done through one pass full tillage by PTOS.

- Tractor operated rotavator was used in farmer's practice for broadcast sowing.
- Data on sowing cost by tractor operated rotavator were collected from different locations.
- Seeds were treated with Provax-200 @ 3 g/kg of seed.
- Fertilizers were applied at the rate of N100+P26+K40+S20+B1 kg/ha.
- Two–thirds of N and total amount of other fertilizers were applied at sowing and remaining N was top dressed after first irrigation.
- Weeding was done at 25-27 DAS by 2-4-D amine @ 1200 ml/ha.
- At maturity, two samples of each plot in 2 m2 for zero till-drill, PTOS, farmer's practices and 2 x 1.1 m for bed planting were harvested.
- After threshed and cleaned, grain yields were recorded at 12% moisture content.
- Data on other plant characteristics during harvesting from each of the experimental plot were recorded, such as spike, plant height, spike length, spikelets/spike, grains/spike, thousand grain wt (TGW), and grain yield were recorded.
- Data were subjected to analysis of variance to sort out significant difference among treatments.
- DMRT was used to compare treatment means.

### 2.6.2 Seed treatment

#### Introduction

Seed treatments control the pathogens that occur on or in the seed. The choice of seed treatment may be dictated by whether the pathogen is borne externally or internally. These pathogens may be important for three reasons:

- Some pathogens do not survive in soil or crop residue and are dependent on the seed borne phase for survival between crops such as fungus that causes loose smut of wheat.
- If a pathogen can survive in soil or residue, being seed borne may allow it to get a head start and, thus, result in more severe disease such as fungus that causes septoria leaf blotch of wheat.
- Seed borne pathogens may hitch a ride to new localities in seed shipments.

#### Description

Loose smut:

- Treat the seed with Tebuconazole @ 1g/kg seed or Vitavax @ 2 g/kg or Bavistin @ 2.5 g/kg seed for the control of loose smut.
- Seed treatment should be done with seed treating drum.
- Seed treatment should also be done with T. viride @ 4 g / Kg seed in combination with carboxin (Vitavax 75 WP) @ 1.25 g / Kg seed or Tebuconazole (Raxil 2 DS) @ 1.0 g / Kg seed.

- ❖ Seed treatment with T. viride alone @ 4 g / Kg seed is also helpful as it reduces the rust severity.

Flag smut:

- ❖ To control flag smut, treat the seed before sowing with Thiram @ 3 g/kg or Tebuconazole 1g/kg or Vitavax @ 2 g or Bavistin @ 2.5 g/kg seed.

Other diseases:

- ❖ Treat the seed with Captan or Thiram @ 3 g/kg seed for the control of root rot, foot rot, seedling blight, black tip and black spot of glumes.
- ❖ Captan and Thiram treatment should not be done earlier than one month of sowing as it affects seed germination.
- ❖ Treat seed before sowing with Trichoderma viride @4 g/kg seed.
- ❖ It provides induced systemic resistance, thus reducing the severity of the yellow rust at later growth stages of the crop
- ❖ In termite prone areas, seed treatment with chlorpyriphos @ 0.9g a.i or Endosulfan @ 2.4g a.i/kg seed, be taken up for their management.
- ❖ Seed treatment with thiamethoxam 70WS (Cruiser 70WS) @ 0.7 g a.i./kg seed or Fipronil (Regent 5FS @ 0.3 g a.i./kg seed) is also very effective
- ❖ The IPM module involves the seed treatment with T. viride (@4g/kg seed) + carboxin (75WP ) @1.25g/kg seed) or Tebuconazole (@ 1.0g/kg seed) for the control of loose smut, followed by broadcast of insecticide treated soil (with endosulfan 35 EC @ 2.3L or chlorpyriphos @ 3L/ha) at 15 DAS for termites.

Ear cockle:

- ❖ The infested seed lot should be floated in 2 percent brine solution for this purpose
- ❖ The galls will float on the surface.
- ❖ These should be separated and destroyed away from the field by burning.
- ❖ The seed should be thoroughly washed to remove the salt solution before sowing.

**Methods**

- ❖ Due to infection in seeds, wheat crop gets different diseases like: smut, rust, blight etc. Due to these diseases, there is adverse effect on annual production of wheat.
- ❖ The infection from such diseases could be minimized or reduced by using only healthy, and quality seeds chosen through hot water seed treatment and treatment with fungicide, which is described detail below:

(A) Hot water seed treatment

- ❖ Initially, soak the seeds in ordinary water for about 4 hours, than dip in hot water 52 °C for 10 minutes or 55 °C for 5 minutes.
- ❖ After that, cool and dry the seeds in shade, and let for fungicide treatment.

(B) Fungicides treatment of seed

- ❖ It is compulsory to treat the seeds for producing quality and healthy wheat crop.

- Mix the wheat seeds before sowing with Bevistin or Benlate2 grams (0.2%) per kg seeds in an air tight pot and stir it vigorously for 8-10 minutes.
- Treating the seeds this way helps in controlling diseases and germs on outer surface of wheat grain such as blast, loose smut, etc.

(C) Seed Dressing

- The most common method of seed treatment.
- The seed is either dressed with a dry formulation or wet treated with a slurry or liquid formulation.
- Dressings are applied both on-farm or in specialised seed treatment facilities.

(D) Seed Coating

- A special binder is used with a formulation to enhance adherence to the seed and begin to impact seed size and shape.
- Coatings require advanced treatment application technology.

(E) Seed Pelleting

- The most sophisticated seed treatment technology, resulting in changing the physical shape of a seed to enhance plantability and handling.
- Pelleting requires specialised application machinery and techniques and is the most expensive of the applications.

## 2.7 NPK requirement in nursery bed

### 2.7.1 Nutrition required in Nursery

All plants/seeds require nutrients to survive and grow and they take nutrients from air, soil, and water, since they are colourless gases or like dust that is dissolved in water or stuck on each piece of soil.

The quantity of nutrients available to nursery plants/seeds is affected by:

- substrate quality
- water quality
- plant type.

Features

- Nutrients are taken up by fine root hairs, not by the big roots.
- Even very large trees have many small, fine root hairs to absorb the nutrients and water they need.
- The larger roots are used for supporting the tree and for storage of water and other plant food.
- The root hairs excrete liquids that affect the acidity of soil (pH) and when pH changes, amount of nutrients present also changes.
- There are many plant food nutrients that are essential for proper seed development.

- ❖ Each of these nutrients are equally important to the plant, but each required in large quantities.
- ❖ The primary (macro) nutrients such as nitrogen, phosphorus, and potassium are most frequently required in nursery.
- ❖ The deficiency due toi primary nutrients will be expressed by seedlings through deficiency symptoms, so required fertilizer needs to be applied as required.
- ❖ In case of general nutrient support, FYM and compost are mostly used in nurseries where available soils are either too sandy or too heavy with low quality as it will helps in building good soil structure, improves water holding capacity of soil and provides nutrients for plants.
- ❖ Such nutrients greatly reduce the need for chemical fertilizers and if mixed with small amounts, dilute the fertilizer, making it available in much larger useful quantities.
- ❖ Gardener should therefore economically use required chemical fertilizers and natural fertilizer.
- ❖ The leaves of seedlings planted with unfinished compost usually turn yellow as plant cannot get all nutrients as it needs whilst the immature compost that continues to absorb what little nitrogen what is available.
- ❖ Adding decomposed manure in nursery mixture will assure of production quality and healthy seedlings.
- ❖ To boost rhizome growth, phosphorous should be added while application of urea will resulted foe good vegetative/foliar growth of seedlings.

## 2.7.2 Types of Nutrients

There are two types of nutrients:

- ❖ Macronutrients: required in large quantities
- ❖ Micronutrients: required in small quantities

| macronutrients | micronutrients |
|---|---|
| Nitrogen (N) | Iron (Fe) |
| Phosphorus (P) | Manganese (Mn) |
| Potassium (K) | Zinc (Zn) |
| Calcium (Ca) | Copper (Cu) |
| Magnesium (Mg) | Boron (B) |
| Sulphur (S) | Chlorine (Cl) |
| | Molybdenum (Mo) |

Nitrogen, phosphorus and potassium, together comprise over 75% of mineral nutrients found in plant.

Nitrogen (N)

- ❖ It has similar benefits as protein.

- It is important for good stalk growth by excess decreases immunity and fruit ability.

Phosphorus (P)

- It strengthens the root systems.
- It has a capacity for seed creation.
- It is disease resistance and pest prevention.
- It improves flowers and blooming.
- It strengthens tissues and flavours in edibles and veggies.

Potassium (K)

- It is important for roots and seed production.
- It supports the ability to tolerate extreme temperatures.
- It helps in extra prevention absorption of other minerals, like magnesium and calcium.

Calcium

- It strengthens and fortifies overall tissues everywhere.
- It helps in neutralize acidity both in plant and in surrounding soil for optimal health.

Magnesium

- It helps in increasing intake of phosphorus and boosts production of chlorophyll.
- It gives healthy green color and encourages absorption of CO2.
- Plants which do have magnesium will have poor colouring and anemic appearance.

While plants need all of these nutrients for optimal health, higher numbers of some are more necessary at some times than others.

### 2.7.3 NPK requirement in Garlic

In garlic

- use of special fertilizers like NPK in ratio of 19:19:19 at 1 % with 30,45 and 60 DAP should be followed by special fertliser NPK at 13:0:45 at 1 % with 75, 90 and 105 DAP in order to increase the yield and to improve storability.
- fertigation in 10 splits are required for higher bulb yield.
- grown on Avid system, integrated nutrient management is required with saving of 50% chemical fertilizers to 70% chemical fertilizers without any significant reduction in yield with improvement in soil health and environmental soundness with standard check.
- farm application of GA3 at 25 ppm with 45 & 60 DAP are required to improve the yield and quality under certain conditions.

## 2.8 Seedling size, age and stage for transplanting

### 2.8.1 Transplanting

- Transplanting is a method of planting seeds in seed box which on bigger seedlings gets transplanted from seed box to field plot.

- It is done when seedlings attains 12 -15 cm height with development of 3 – 4 leaves.
- Transplanting is planting with the use of pre-grown seedlings or plants that had been propagated from seeds.
- Crop area is the space which is particularly used for growing crops.
- It is indirect as seeds are not immediately sown, but are first used to raise the seedlings in pots or in temporary place which are further used when reaches at right age.
- Field planting is done either manually or by mechanical means.

Direct seeding or transplanting depends on:

- crop species to be grown
- ease in planting and survival rate
- gardener's familiarity
- timeliness
- financial capability of gardener/owner
- return on investment

### 2.8.2 Age of seedling

The seedling age at transplanting has direct bearing on yield of crop where seedling has 3-4 leaves. The optimum age of seed depends on:

- duration of variety and season
- date of planting
- system of raising nursery
- tillering habit
- inherent fertility
- fertilizer management
- cultural practices

Age of seedling for timely planting

- ❖ Seedlings of 3-4 week old for short duration varieties and 5-6 weeks old for varieties of 5-6 months duration may be considered as optimum age for transplanting.
- ❖ Aged seedlings of 7-11 weeks old, can be used without detriment to yield for longer duration varieties provided, planting is done in optimum period.
- ❖ In short duration varieties, it is essential to use seedlings of optimum age of 3-4 weeks only, irrespective of time of planting, as short duration are bound period.
- ❖ Due to aberrant weather conditions and/or intensive cropping system, transplanting is taken up late beyond critical period.

Age of seedling for late planting

- ❖ Under such situation, economically viable yield will not be possible without adoption of special measures to compensate loss that exists due to delay in planting.
- ❖ Late planting of aged seedlings are better than younger seedlings.
- ❖ Medium and long duration seedling behaves in same way as in case of late planting.

Age of seedling for seasons

- ❖ During certain season, seedlings requires more time to come to plant due to low temperature.
- ❖ The optimum age of seed for planting is comparative more than that for kharif.
- ❖ 30-45 days old seedling performs better than young or old seedling.

### 2.8.3 Seed Size of Garlic

- ❖ Garlic seed is made in abundance after garlic plants flower in late spring or early summer.
- ❖ In late summer and early fall, garlic plants mature into heavy heads with small garlic bulbs which serves as true garlic seed.
- ❖ Small garlic seed/cloves produces small bulbs, in comparison with large seed.
- ❖ Large garlic seed is about 3.5 gm/clove which ranges from 3.0 to 4.0 gm/clove while small garlic seed is around 1.5 gm/clove which ranges from 1.4 to 1.6 gm/clove.
- ❖ Garlic cloves of around 8-10 mm size with spacing of around 12 cm x 8 cm during October is recommended for getting high bulb yield, but for quality bulb production, planting at 15 cm x 10 cm spacing is required.

## 2.9 Kharif onion grown through sets

### 2.9.1 Introduction

- ❖ The bulbs are medium large, deep dark red with average yield potential of 120 q/acre which needs suitable for sowing in kharif season. The best time for sowing kharif onion nursery is middle of June, while to produce bulb sets, is middle of March.

### 2.9.2 Procedure

- To grow kharif onion crop by avoiding failure of seedling during June, plant the crop with bulb sets.
- Keep nursery bed and field area ratio of 1:20.
- Nursery soil should be well prepared and mixed well with rotten 125 kg farmyard manure per marla.
- Sow seed 1 to 2 cm. deep in lines with 5 cm spacing.
- The seed should be sown uniformly in lines, covered with thin layer of well rotten farmyard manure.
- The seed should be sown in proper moisture condition.
- To raise bulb sets, sow 5 kg. seed in beds of 8 marla 200 sq.m. in middle of March.
- Irrigate the seedlings twice a week.
- Medium sized bulb sets of 1.5-2.5 cm diameter gives maximum marketable yield.
- In last week of June, uproot the bulb sets and store in well ventilated dry room in baskets.
- Plant the bulb sets in field in second fortnight of August.
- The bulb crop will be ready for harvesting by the end of November.
- Add 20 tonnes of farmyard manure with 40 kg of N, 20 kg of P2O5 and 20 kg of K2O per acre.
- Use whole farmyard manure, P2O5 and K2O and ½ N before transplanting and remaining dose of nitrogen as top-dressing after 4 weeks of transplanting.

## 2.10 Proper size and quantity of set required per ha (how much)

- Onions are a cool season crop.
- The germination of onions is slow at 6 to 7 C, the optimum germination temperature range is 10 to 35 C, and the maximum temperature is 40 C.
- The onion crop is adapted to a growing season with air temperatures at 13 to 24 C.
- Low temperatures early in the season are desirable with higher temperatures after bulb formation.
- Around 2000 to 4000 kg of sets are required per hectare depending on the set size.
- Approximately 4 kg of raw seed is required for dry onions which are not precision seeded while 6 kg of seed is required for bunching onions and 35-45 kg per hectare is required for pickling onions.
- For growing sets, sow seed 70 to 80 kg per hectare in bands.

## 2.11 Prepare seed bed for plantation

### 2.11.1 Introduction

Seedbed is local soil environment where seeds are planted that has soil along with specially prepared cold frame, hotbed or raised bed which are used to grow seedlings in controlled environment in large young plants before transplanting it into garden or field. Seedbed preparation in farm fields is done by using harrows and cultivators while seedbed preparation in gardens is done by hand tools like rakes and hoes.

### 2.11.2 Features

Preparation of seedbed include:

Removal of debris

- Insect eggs and disease spores are found in plant debris which needs to be removed from the plot.
- Stones and larger debris also physically prevent the seedlings from growing.

Levelling

- The site should be levelled for even drainage.

Breaking of soil

- Compacted soil will be broken by digging which allow air and water to enter that helps seedling to penetrate the soil.
- Smaller seeds needs fine soil structure.
- The surface soil can be broken down in fine granular structure by rake.

Soil improvement

- Soil structure can be improved by introduction of organic matter like compost or peat.

Fertilizing

- Nitrate and phosphate levels of soil can be adjusted with fertilizer.
- If soil is deficient in any micro nutrients, they can be added up.

### 2.11.3 Procedure

- Onion seeds are sown on nursery beds to raise seedlings for transplanting in field.
- Raise the beds with size 3 x 0.6 m and 10-15 cm in height.
- Keep distance of around 70 cm between the two beds for carry out operations such as watering, weeding, etc.

- The surface of beds should be smooth and well levelled.
- Apply 15-20 gm of single super phosphate at the time of bed preparation for each bed.
- Raise the beds such that it avoids problem of water logging in heavy soils.
- The beds should be irrigated for 15-20 days, before sowing and should be covered with 250 gauge transparent polythene for soil solarisation.

- In sandy soils, sowing can be taken in flat beds.
- Seed bulbs should be treated with thiram at rate of 2 gm/kg of seed.
- Sowing of seeds should be done in lines at a proper spacing 5-7 cm row-to-row.
- To avoid mortality of seedlings due to damping off, drenching of beds with Bavistin is effective.
- Seeds after sowing should be covered with fine powdered compost followed by light watering.
- The beds should be covered with dry straw or grass or sugarcane leaves to maintain required temperature and moisture.

## 2.12 Treatment before sowing (sets)

### 2.12.1 Onion Set

Onion sets are small bulblets up to 30-40 mm-diameter produced in high density seedbeds under bulb-inducing day-length. Sets are the easiest of the three planting techniques and a good way to produce a lot of big onions for storage.

### 2.12.2 Seed treatment

- It describes products and processes.
- Seed treatments can be used to control pathogens that occur on or in seeds.
- By treating the bulbs before planting, disease infection can be prevented as all of the fungal inoculum remaining in the seeds will be destroyed by this practice.
- The usages of specific products and specific techniques improve the growth environment for seed, seedlings and young plants.
- It shows complexity ranges from basic dressing to coating and pelleting.
- Due to infection in seeds, crop gets different diseases like: smut, rust, blight etc.
- It is used on many crops to control variety of pests and ensure uniform stand establishment by protecting against soil borne pathogens and insects.
- It can be used to suppress root rots in certain crops.
- It can protect seed and seedling from attack by certain insects and pathogens.
- With such diseases, there is adverse effect on annual production of crop.
- The infection from diseases could be minimized or reduced by using only healthy, and quality seeds chosen through hot water seed treatment and treatment with fungicide.

**Methods of seed treatment**

(A) Hot water seed treatment

- Initially, soak the seeds in ordinary water for about 4 hours, than dip in hot water 52 °C for 10 minutes or 55 °C for 5 minutes.
- After that, cool and dry the seeds in shade, and let for fungicide treatment.

(B) Fungicides treatment of seed

- It is compulsory to treat the seeds for producing quality and healthy crop.
- Mix the wheat seeds before sowing with Bevistin or Benlate 2 grams (0.2%) per kg seeds in an air tight pot and stir it vigorously for 8-10 minutes.
- Treating the seeds this way helps in controlling diseases and germs on outer surface of wheat grain such as blast, loose smut, etc.

(C) Seed Dressing

- The most common method of seed treatment.
- The seed is either dressed with a dry formulation or wet treated with a slurry or liquid formulation.
- Dressings are applied both on-farm or in specialised seed treatment facilities.

(D) Seed Coating

- A special binder is used with a formulation to enhance adherence to the seed and begin to impact seed size and shape.
- Coatings require advanced treatment application technology.

(E) Seed Pelleting

- The most sophisticated seed treatment technology, resulting in changing the physical shape of a seed to enhance plantability and handling.
- Pelleting requires specialised application machinery and techniques and is the most expensive of the applications.

## 2.13 Identify the most suitable time for planting bulb crop

- Quality of seed depends on season in which it is grown.
- The time of planting has great impact over seed yield and incidence of disease.
- Onion seed production requires two different seasons where in one season, quality bulbs are produced and are again planted while in second season, true seed production occurs.
- First season is good for healthy and quality bulb production while second season is suitable for better growth of bulb so that it can facilitate good flowering and seed setting.
- June-September is the best season for bulb production while October-March is good season for true seed production.
- Low temperature prevailing in early second season facilitates easy flowering of bolting

while dry weather, medium RH and temperatures are important for seed setting and maturity.

- ❖ If late planting is done, vegetative growth is less and due to this there are lesser number of flowers and also the seeds in umbel and attack of thrips is also more.

## 2.14 Planting bulb crop with appropriate method

- ❖ There are two methods of seed production.

1. Seed to seed method
2. Bulb to seed method

Seed to Seed method

- ❖ In this, seedlings are transplanted in late kharif which on over wintering allows flowering.
- ❖ The method does not allow examining of mature bulb features and seed characters.
- ❖ The method is not popular since all varieties are not good for annual seed production due to poor bolting ability and lower seed yield.
- ❖ The seeds produced with such method are not good for further multiplication.

Bulb to Seed method

- ❖ In this, seeds are sown in kharif season.
- ❖ When seedlings are 12-15 cm in height and are 45-50 days, gets transplanted to main field for raising healthy bulb crop.
- ❖ The bulbs are harvested when 75% plants show neck fall or top die down.
- ❖ Once curing, damaged, twin bulbs and long necked bulbs gets discarded.
- ❖ The medium and large sized bulbs are selected and stored known as mother bulbs.
- ❖ The optimum temperature about 12 C is good for storage of mother bulbs.
- ❖ The plants from mother bulbs produce early and higher seed yield.

## 2.15 Perform plantation of bulb crop at proper spacing (R-R and P-P)

Spacing shows plant density which depends on expected growth of particular crop plant variety in given agro-climatic region. The optimum plant population is an important factors for optimum utilization of solar energy and soil nutrients to increase the yield per hectare of the bulb crop, where only single underground bulb is produced per plant. Higher plant population can be achieved by lowering the distance between two rows or between two plants in the row, so use of proper geometry to get appropriate plant stand is pre-requisite for high crop yield per unit area.

Features

- ❖ Different spacing in onion transplanting affects plant height, length of leaf, number of leaf, bulb length, bulb diameter, multiplier bulb and yield per ha.
- ❖ Closer spacing of around 7.5x7.5 cm produces high yield of around 16.65 t/ha which is followed by yield of 15.82 t/ha with spacing of around 10x7.5 cm.

- The lowest yield of 10.05 t/ha was obtained from wider spacing of 20x10 cm.
- The highest multiplier bulbs with 24.34% from wide spacing of 20x10 cm follows spacing of around 15x10 cm.
- The lowest percentage of multiplier bulbs with 14.38% from closer spacing of 7.5x7.5 cm. 12.42 and 12.97 t/ha yield can be obtained from 15x10 and 15x7.5 cm spacing.
- The highest plant height of 40.46 cm can be obtained from spacing of 20x10 cm which follows spacing of 15x10 cm.
- The shortest plant height of around 27.07 cm can be seen from 7.5x7.5 cm spacing.
- The heaviest bulb of 23.52 g can be produced by spacing of around 20x10 cm which is followed by spacing of 15x10 cm with lowest spacing of 7.5x7.5 cm.
- The biggest bulb obtained from spacing of 20x10 cm is followed by spacing of 15 x 10 cm and lowest by spacing of 7.5 x 7.5 cm.

## 2.16 Transplanting of seedling in main fields

- Transplanting healthy and vigorous seedlings in main field can be useful for seedling to grow faster by using applied nutrients to develop resistance to pest and diseases and to avoid gap filling operations.
- The healthy seedlings pull out from maintained nursery can be used for transplanting.
- After irrigation, complete soaking of ridges, single seedling can be transplanted at spacing of 10 cm.
- When seedling is about 6-8 weeks old and 15cm high, they should be removed from nursery beds and transplanted into main field at a distance of 15-20cm between and 7-10cm between plants or seedling.
- After this, grower should irrigate the field soon.
- Transplanting should be done either in early hours or late in evening to avoid drying of seeds.

# Chapter 3

# Soil Nutrient Management in Bulb Crops

## 3.1 Select appropriate organic fertilizer including farm yard manure for its application

### 3.1.1 Fertilizer

Fertilizer is a substance added to the soil to improve its fertility. Bulb crop grower should take care of plants by giving proper fertilizer. He/she should properly understand, when fertilizer should be used. Fertilizers supply plants with three major elements such as:

Nitrogen

- Promotes vigorous blade, leaf and stem growth
- Encourages dark green colour
- Assists in chlorophyll production
- Improves overall plant health, quality, and density
- Aids in tolerance to heat, cold, and drought stress

Phosphorus

- Promotes early plant development
- Hastens plant maturity
- Promotes bloom, fruit, and root development

Potassium

- Improves tolerance to drought, disease and heat
- Increases winter hardiness
- Aids in water uptake
- Enhances overall health and vigor
- Fertilizer in small quantities contains secondary nutrients like calcium, magnesium and sulphur. Fertilizers rich in nitrogen will stimulate healthy, green growth of foliage, while those rich in phosphorus will encourage good root development and improved flowering. Also, fertilizer rich in potassium will help in building up reserves for plants that have a dormant period.

Fertilizer comes in granular slow release form or compressed spike that is pushed in soil. It is also available as powder or liquid that is diluted with water.

### 3.1.2 Time of application

- Spring is best time to fertilize when outdoor plants start growing again.
- There is time when fertilizers release nutrients, so bulb crop grower should know when to re-apply.
- Many granular fertilizer is slow-release which can be applied only once or twice a year for perennial and mixed borders.
- Lawns can be fertilized 2-4 times a year as per general condition, as fertilizer spikes are put in ground once in spring.

- Anything mixed with water is easily used by plants so it can be applied more often.
- Grass needs fall feeding, so it is better to stop fertilizing after September and rose plant by August for natural cycle of dormancy instead of forcing growth.

### 3.1.3 Types of Fertilizer

After establishing that your plant needs to be fertilized, it is necessary to decide what type of fertilizer to use. Fertilizers come in several different varieties:

- Organic
- Inorganic
- Chemical
- Foliar

**Organic Fertilizer**

- It is composed of naturally occurring biodegradable materials.
- Many organic fertilizers are made with animal manure plus, compost, seaweed, peat moss, mineral deposits and other ingredients from nature.
- Such fertilizers are good for gardens.
- It helps in increasing direct yield of plants once fertilize.
- It improves the health and long-term productivity of the soil.

**Inorganic Fertilizer**

- It comes in different forms such as liquid, powder and granular.
- It is made from concentrated ammonia that is diluted with water.
- Rock phosphate and potassium are many added to make a compound fertilizer.

- Such fertilizers are used in treatment of sizable industrial fields as they are cheaper and more easily produced on large scale.
- It is less bulky than organic fertilizers as it allows the plant to carry out nutrients more easily from the soil to its leaves and fruits.
- It lead to trace mineral depletion over time as many such fertilizers do not put minerals back into ground what organic fertilizers do and have lower mineral content in fruit and vegetables.

**Chemical Fertilizer**

- Such fertilizers are inexpensive and widespread.
- In this, rapid-growing plants takes up nutrients fast unless the formula is designed to be time-release.
- It results in risk of over-application which causes burning and absence of soil-improving qualities.
- The fertilizers come in the range of formulations such as pellets, liquid concentrates and powders.
- Such fertilizers come in pre-measured packets designed which added to watering.
- These fertilizers make it convenient for gardener to apply fertilizers to containers, houseplants, or landscape.

**Foliar Fertilizer**

- These are liquid nutrients that plants absorbs by their leaves.
- Not all flowers feed efficiently this way as waxes and hairs on leaves act as a barrier to nutrient uptake.
- Flowering plants cannot receive all nutrients they require from their leaves, but there results some nutrient deficiencies quickly by using such fertilizers.
- Potassium is one readily absorbed nutrient in foliar feeding applications, hence using foliar fertilizers in flower garden will show potassium deficiencies.
- In case of flowers showing signs of exhibiting of chlorosis, then the plant is suffering from iron deficiency.

**Special Purpose Fertilizer**

- These are formulated for specific plant and soil needs.
- They are grouped to stimulate growth or production.
- For stimulating growth, there are special fertilizers for specific seasons having large nitrogen concentration with N-P-K ratios as 16-6-4.
- For stimulating production, there are special fertilizers which contains low doses of nitrogen with N-P-K ratio as 3-20-20.
- The fertilizer with N-P-K ratio as 3-20-20 foor bes results should be applied in dry or liquid form to plants when established.
- Plant-Specific fertilizers boost N-P-K ratios for properly nourish and premium results from particular plant and are useful for citrus trees and acid-loving plants

### 3.1.4 Organic Fertilizer

- It comprises of variety of plant-derived materials that ranges from fresh or dried plant material to animal manures and litters to agricultural by-products.
- The nutrient content of organic fertilizers varies greatly among source materials, and readily biodegradable materials make better nutrient sources.
- Nitrogen and phosphorus content is lower, often substantially lower, in organic fertilizers compared to chemical fertilizers.
- Moisture content is another factor that reduces or dilutes the nitrogen and phosphorus concentrations of organic fertilizers.
- It provides energy for biological activity.
- Many of the nutrients used by plants are held in organic matter until the organisms decompose the materials and release them for the plants' use.
- It attracts and holds plant nutrients in an available state, reducing the amount of nutrients lost through leaching.
- It improves soil structure, so that air reaches plant roots and the soil retains moisture.
- The organic matter and the organisms that feed on it are central to the nutrient cycle.

### 3.1.5 Manure

#### For Garlic

- While adding manure, bulb crop grower should not add manure too close to harvest time. Organic certification usually requires applying it no sooner than 120 days before harvest. Antibiotics, drugs, and GMO residue from animal feed take time to break down in the manure.

**Cow Manure**

- Three weeks prior to planting, bulb crop grower needs to have their fields sprayed with liquid cow manure.
- Spraying the field includes planting and field that will be planted in the following year.
- In three weeks, smell has dissipated enough that planting is not an odious chore, and it also allows nitrogen to settle so that it doesn't shock the cloves when planted.
- Bulb crop grower should be sure when to turn under the liquid manure after spraying so that nutrients don't evaporate.

**Poultry Manure**

- Poultry manure like chicken manure is most concentrated source of nitrogen of any manure.
- Buying or using organic manure is best if bulb crop grower can manage as there are so many chemicals and GMOs used in feed these days.
- Poultry manure doesn't evaporate like cow manure, so it doesn't need to be turned under as quickly as cow manure, but it's still important to mix it into the soil before planting especially since that nitrogen is so concentrated.

**Horse Manure**

- Composted horse manure can be a good source of nitrogen.
- Bulb crop grower should make sure that if any drugs have been given to horses, will typically be shed in the manure which goes straight into the soil.

## 3.2 Apply organic and inorganic fertilizer in correct dosage and approximate time

### Determine right organic/inorganic fertilizers to be applied

- Inorganic fertilizers in combination with organic fertilizers help in improving soil health and sustain crop yield.
- For better crop production, both organic and inorganic plays an important role.
- Fertilizer use, though has been a prime contributing factor for the increased agricultural productivity in the areas, there are situations when they are not available in right time and adequate quantity despite high priced.
- After the soil test results, wheat cultivator can be able to judge the required supplemental fertilizer along with type of fertilizer to be used.

The factors which involves in selecting correct fertilizer include:

- physical properties
- chemical properties
- environmental stewardship
- economics

The good fertilizer selection involves:

- Solids, liquids or gases forms where each physical form has its own uses and limitations that provides the basis for selecting the best fertilizer for the field.
- Correct solubility, as it shows how readily the nutrients are dissolved in soil water and are taken by the plants.
- Required salt index as soluble salts at high concentrations in soil solution will injure/ kill plants or prevent germination of seeds.
- Checking of acidity of some fertilizer materials.
- Correct particle size as it is important for producing stable, high-quality blended fertilizer.
- Comparison of material price per actual plant food by formula, (Price per ton of fertilizer)/(2000 x plant food content as decimal value) = per rupee of plant food

**Apply fertilizers in right dosage, time and intervals of crop growth**

- Timing of fertilizer application has a significant effect on crop yields.
- Proper timing of the fertilizer application increases yields, reduces nutrient losses, increases nutrient use efficiency and prevents damage to the environment.
- Applying fertilizers at wrong time might result in nutrient losses, waste of fertilizer and even damage to the crop.

- Plants need different nutrient rates and ratios at different growth stages.
- To make the nutrients available when plant needs them, fertilizers should be applied at the right timing.
- The optimum timing for fertilizer application is determined by Nutrient Uptake Pattern of the crop.
- For similar crop, each nutrient has an individual uptake pattern.
- The maximum rate of fertilizer that can be applied at one application depends on the salinity threshold that the crop can tolerate.
- Split fertilizer applications helps in avoiding salt damages to the crop and improving germination rate.
- Soil type affects the timing and frequency of fertilizer application where Cation Exchange Capacity and Soil Texture properties determine the frequency and timing of application.
- If nitrogen is applied too early before need of plant, then a required portion may get lost before the crop takes it up, so time nitrogen is in soil before the plant takes up should be minimized.

## 3.3 Apply vermi compost and interaction with its expert

### 3.3.1 Introduction

- Compost is produced by aerobic (oxygen requiring) microorganisms activity as such microbes require oxygen, moisture and food to grow and multiply.
- Vermicompost is the excrement which is left behind by earthworms after they digest decomposing organic matter.
- Bulb Crop Growers can harvest vermicompost and use it as a soil amendment, best known for increasing crop yields and improving drought tolerance in plants.
- For making correct vermin compost, many vermiculture experts recommend one-to-one ratio as approximately 1,000 worms to one pound of garbage to be added.
- As organic fertilizer, vermicompost serves as substitute for synthetic fertilizer in soil-enriched vegetable production.

### 3.3.2 Applying Vermicompost

- Bulb crop grower can use vermicompost immediately or can store it for up to year and use it throughout the crop season.
- Worm castings can be directly mixed with potting soil or field soil as organic fertilizer.
- The compost can be used as top dressing for indoor or outdoor crops/plants.

Use like compost:

- Dig the compost lightly in topsoil when mulching grow beds or preparing new beds or in case of applying light top dressing around mature plants, shrubs, and trees.
- Vermicompost goes about three times as far as ordinary compost, so crop grower should use one-third as much.

- Crop grower should use 3 tablespoons of vermicompost per plant and then water each plant by taking care to soak the worm compost which has just spread so that it can work with nutrients down in soil.

Use for germinating seeds:

- Crop grower should cast mixed 50/50 with potting soil to produce growing medium for seedlings which gives plant a great start to life.

Use on bare-root transplants:

- Spray the roots with diluted worm tea solution to help preventing transplant shock.
- Storing:
- Crop grower should keep the compost in breathable container where castings will not dry out.
- In such case, the active microbes can be kept for upto an year.

Advantages

- Easily done inside the home
- Clean, odor free true only if process is well managed
- On adding to soil, will increases the organic content and adds nutrients
- Improves soil porosity
- Improves moisture holding capacity of soil
- Provides slow release nutrients
- Will not burn like synthetic fertilizers

Pros

- Easily done in small batches.
- Works during cold weather if done indoors.
- Less nitrogen loss.

Cons

- Crop grower has to buy and deal with the worms
- Worms have to be separated from compost before it is added to field.
- Not easily managed by owners for large amounts of yard waste.
- Does not kill weed seeds.
- Produces nitrous oxide and methane which is harmful greenhouse gases.
- Leachate may spread plant pathogens.

## 3.4 NPK requirement of bulb crop as par package & practices for Haryana

### 3.4.1 Introduction

- The different types of chemical fertilizers are classified as per principal elements, Nitrogen (N), Phosphorous (P) and Potassium (K), which may therefore, included in more than one group.

- The combinations of NPK significantly affected most important bulb characteristics
- The essential primary macro nutrients like nitrogen, phosphorus and potassium (NPK) are necessary for onion plant growth, maturity, bulb yield and quality.

### 3.4.2 For Garlic:

- Nitrogen: Garlic needs more nitrogen than most farmers think, mostly during its initial growth phase as it emerges and spreads its leaves. Adding organic manures, such as cow and poultry, are a great way to add nitrogen.
- Phosphorus: Phosphorus is needed for optimal root development.
- Potassium: Enough potassium is critical for leaf growth and healthy bulb formation.

### 3.4.3 For Onion

- Onion is a long season crop having special root architecture as shallow root system, low root densities and lack root hairs, so it requires high concentration of N, P and K in soil solution to satisfy potential demand, hence it needs for high levels of N, P and K in the soil.
- As nitrogen is main part of chlorophyll, it is essential for synthesis of proteins, enzymes and promotes vigorous vegetative growth while phosphorus and potassium are used for photosynthesis, respiration, energy storage (ATP, ADP formation) and in enhancing translocation of assimilates and protein synthesis, so for economical feasibility of onion, application of NPK fertilization in balanced ratio is prerequisite.

## 3.5 Method of fertilizer application

### 3.5.1 Forms of Fertilizer Application

#### 1. Solid Form

There are different methods of fertilizer application in solid form such as:

Broadcasting

- It refers to spreading fertilizers uniformly all over the field by hand.
- Suitable for crops with dense stand, the plant roots permeate the whole volume of the soil, large doses of fertilizers are applied and insoluble phosphatic fertilizers such as rock phosphate are used.
- The fertilizers are generally applied to the soil with the last preparatory tillage just before planting.

(a) Broadcasting at planting:

- The fertiliser is broadcasted just before planting, at the time of ploughing.
- Phosphoric and potassic fertiliser and half of nitrogenous fertiliser as per recommended dose for a particular crop is applied by broadcasting during planting.

(b) Top dressing:

- The method of application of fertiliser in the standing crops is known as top dressing.
- It provide nutrients such as nitrogen, for the growth of plants.

- ❖ Topdressing of phosphatic and potassic fertilisers is ordinarily done on pastureland and in orchards or only on fruit trees growing in between agricultural crops.
- ❖ Topdressing with nitrogenous and potassic fertiliser should not be done when the leaves of plants are wet which burns or scorch the leaves if applied in morning before 8 a.m. or after rain.

Placement

- ❖ It refers to the placement of fertilizers in soil at a specific place with or without reference to the position of the seed.
- ❖ Placement of fertilizers is normally recommended when the quantity of fertilizers to apply is small, development of the root system is poor, soil have a low level of fertility and to apply phosphatic and potassic fertilizer.

(a) Plough sole placement:

- ❖ The fertiliser is placed in the plough sole after opening the furrow with the plough, and this furrow is covered immediately as the next furrow is turned.
- ❖ This method has been recommended in dry soil where there is moisture only in the plough-sole layer and in problem soils where there is a problem of fixation.

(b) Deep placement:

- ❖ The ammoniacal nitrogenous fertilizer or ammonia forming nitrogenous fertiliser is placed deep in the reduced layer to check denitrification.

(c) Sub-soil placement:

- ❖ The phosphatic and potassium fertilisers are placed in sub-soil with the help of heavy machinery to avoid their fixation in strongly acidic soil.
- ❖ This method is recommended in humid and sub-humid region.

Band placement

- ❖ In this method, the fertilizer is placed in bands on one side or both sides of row, about 5 cms. away from the seed or plant in any direction.

Band placement is of two types.

i) Hill placement: It is practiced for the application of fertilizers in orchards. In this method, fertilizers are placed close to the plant in bands on one or both sides of the plant. The length and depth of the band varies with the nature of the crop.

ii) Row placement: When the crops like sugarcane, potato, maize, cereals etc., are sown close together in rows, the fertilizer is applied in continuous bands on one or both sides of the row, which is known as row placement.

d) Pellet application

- ❖ It refers to the placement of nitrogenous fertilizer in the form of pellets 2.5 to 5 cm deep between the rows of the paddy crop.
- ❖ The fertilizer is mixed with the soil in the ratio of 1:10 and made small pellets of convenient size to deposit in the mud of paddy fields.
- ❖ The pellets are deposited in the soft mud of paddy fields.

### 2. Liquid form

Liquid form of fertilizers are applied by the following methods:

- Starter solution: Starter solution is prepared by mixing N, P, and K fertilizers in the ratio of 1 :2 :1 or 1 :1:2. This is applied to vegetable seedlings at the time of transplating and it helps rapid establishment and quick growth of seedlings.
- Foliar application: This is a method of spraying the leaves, of growing plants, with suitable fertilizer solution of a low concentration
- Direct application to soil: Liquid fertilizers such as anhydrous ammonia and nitrogen solution are directly applied to the soil with special injection equipment. Liquid manures such as urine, sewage water and cowshed washing are let into the field.
- Application through irrigation water: Fertilizers, both straight and mixed fertilisers, are easily soluble in water. They are allowed to dissolve in the irrigation water and applied to the soil through irrigation. Nitrogenous fertilizers are, generally, applied through irrigation water.

## 3.6 Stages of application of fertilizer

- Fertilization of onion plants is vital to success as onion growth and yield will increase by banding phosphorus 2-3 inches below seed at planting time.
- The phosphorus acts as starter solution which invigorates the growth of young seedlings.
- Banding phosphorus like super phosphate (0-20-0), 2-3 inches below the seed involves making a trench of 3 inches deep which distributes 1/2 cup of super phosphate per 10 row feet, covering phosphate with soil, sowing seed and covering lightly with 1/2 inch or less of soil.
- Once established, onion plants should receive additional amounts of fertilizer (21-0-0 Ammonium sulfate or Ammonium nitrate) as side-dress application every month.
- Onions require a high source of nitrogen.
- A nitrogen-based fertilizer such as ammonium sulfate or ammonium nitrate should be applied at rate of one cup per twenty feet of row.
- Application of ammonia should be done about three weeks after planting and then continue every 2 to 3 weeks.
- Once the neck starts feeling soft, then crop grower should not apply more fertilizer as it should take place 4 weeks before harvest.
- Crop grower should always water the field immediately after feeding and maintaining moisture during growing season, as closer harvesting of onion requires more water.

## 3.7 Application of foliar spray of fertilization

- Foliar application is used to supply fertilizers to developing crops which provides flexibility to growers in their application.
- Foliar fertilizer sprays are used to improve quality and yield of bulb crop.

- There are two different types of foliar fertilizers which were used for bulb crops where one is high nitrogen fertilizer made of hydrolyzed fish powder with 12% N, 0.25% P, and 1.0% K having high solubility while other foliar fertilizer is high in potassium extracted from sea kelp that has 70+ growth hormones, cytokinins, auxins, vitamins, and enzymes which is rich in trace elements with 1% N, 0% P and 4% K.
- The plants were sprayed with the foliar feed to the point of run-off early in morning or in late evening when sun was not as intense and there was no wind.
- The amount of mineral nutrients available in foliar sprays is very small when seen against normal nutrient needs of plants.
- Such mineral nutrients promote usage of sprays which emphasizes importance of timing of spray application, which shows that foliar nutrients can be helpful if applied at critical point in life cycle of crop/plant.
- Spring fertilizer application is normal procedure on fields which was made at time that shows half of plots treated with foliar sprays which was applied as top dressing at rates shown.
- Among fertilizer application methods, application of foliar nutrition is used as it facilitate easy and quick consumption of nutrients by penetrating stomata or leaf cuticle that enters the cells.
- Spraying onion plants with Fe or Zn at 60-70 days after transplanting date leds to significant increase of plant vegetative growth, bulb yield and quality.

## 3.8 Application of sulfur fertilizer in bulb crop

- Apart from NPK nutrients, sulphur is also an essential plant nutrient which is important for onion crop for improving yield and pungency of onion bulbs.
- Sulphur increases bulb yield of onion and improves its quality, especially pungency and flavour.
- Sulphur is recommended as basal dose during the time of transplanting for bulb crops.
- The application of 15 kg sulphur/ha is good for growing onion crops in soils having sulphur level above 25 kg/ha while 30 kg sulphur/ha is required for soils having sulphur level below 25 kg/ha for optimum bulb crop production.
- Soil application of 50 kg S/ha is recommended for long day bulb crops.
- Original source of soil sulphur is metal sulphide minerals which when exposed to weathering, S-2 oxidizes to SO4-2 and uptake of sulfur into roots from the soil is exclusively as sulfate uptake.
- Sulphur deficient crop/plants had poor utilization of macro and micronutrients which significantly lower total solids in onion bulbs at maturity.
- Sulfur compounds are directly related to garlic's unique healing benefits and flavors.
- To add sulfur, bulb grower has to sprinkle gypsum over beds in the spring, after the plants have emerged and begun to leaf out.

## 3.9 Maintain the record of fertilizer application

### 3.9.1 Records

Keeping track of what to buy and knowing happening on fields requires some records. Good field management requires good useful set of field records which does not ensure successful of fields, but success is unlikely without them.

Record-keeping is a task that has been started many times by eager farmers/growers, only to be dropped later. Reasons that complete, accurate records are not kept revolve around lack of time and motivation. Keeping records is productive work.

Good record-keeping system should be taken into account for seed growers as per needs, abilities and limitations. Seed growers should need to understand the purpose for keeping the records and when the purpose is known, it will help them in deciding which records are useful and need to be kept.

With accurate facts and figures is useful while borrowing money, seeking government support and completing tax returns. From reliable farm records, equity proportion of total assets owned can be updated to assist with future farm investment programs.

Field records can help seed growers in evaluating seed enterprise on mixed field while field advice will be more effective if adviser knows exactly what is happening on the field. Records and statements show what has occurred or is occurring on the field while budgets predict what might happen in the future.

Records provide seed grower with data, information and knowledge. There are four uses for field records such as:

- ❖ Use as service tool which covers income tax calculations, estate planning, business arrangement reconciliation and obtaining and managing credit.
- ❖ Providing data for financial analysis and diagnostic instruments like identifying strengths and weaknesses of business.
- ❖ Progress indicator where series of records are required to monitor work progress.
- ❖ Forward planning tool for keeping past records to project future cash flows for current and modified farm management practices.

### 3.9.2 Details of Record

Records of field comprises of:

Field diary

- ❖ It is a form of small notebook where seed grower should record key facts and figures about farm business and its day-to-day activities.
- ❖ Farmers /seed grower who uses diary finds important facts and figures which could be easily lost or forgotten are stored permanently in this for future reference.

Seed Crop records

It is useful to record what happens to each plot or seed crop each year that will includes:

- ❖ crop,
- ❖ fertilizer applications,

- agronomy (seeding, weeding, sprays),
- visual impression of crop,
- harvest interval of forage crops
- seed crop yield

| ADDRESS OR LOCATION OF THE APPLICATION SITE | Name of Applicator | RATE APPLIED | NUTRIENT VALUE OF THE FERTILIZER MATERIAL | TYPE OF FERTILIZER MATERIAL APPLIED | METHOD OF FERTILIZER APPLICATION | DATE OF APPLICATION |
|---|---|---|---|---|---|---|
| | | | | | | |
| | | | | | | |
| | | | | | | |
| | | | | | | |
| | | | | | | |
| | | | | | | |
| | | | | | | |
| | | | | | | |
| | | | | | | |
| | | | | | | |
| | | | | | | |
| | | | | | | |

Unit costs

It covers all main field inputs like:

- fertilizers,
- fuel,
- irrigation water,
- concentrates and their ingredients,
- purchased forages,
- stock purchases

The details are required for routine book keeping and in monitoring seasonal changes and will help in planning future purchases.

Unit returns

- It includes returns from all field outputs such as sale of crop, manure etc.

Crop stock accounting

- The main purpose of seed crop stock accounts is to monitor net losses and gains in income, and to distinguish the increases and decreases due to changes in market value.
- When quantifying total stock value, seed grower has to take into account both in change size and change in unit of the crop.

Plant and improvement records

It covers records related to:

- date and cost of purchase or installation,

- annual depreciation,
- insurance and registration,
- fuel use,
- hours use,
- repairs and maintenance

Key financial records

- It covers detail related to interest and principal repayment schedules to plan repayments.

Other Field Charges

It include records related to:

- rates and government charges,
- telephone and office upkeep charges,
- labourer's wages,
- capital investments,
- unusual weather events,
- important meetings,
- credit repayments,
- changes in valuation of capital

Personal expenses

- It includes expenses related to personal input wages which includes valuation of any farm produce used for personal consumption.

Different records

- It covers cost related to crops, fertilizer, seeds, spray, labour, water, machinery, harvesting, processing, transport and selling.
- In this, wages are broken down into different farm activities.

### 3.9.3 Importance of records

- They are management tool which allows seed growers to measure how efficiently they are using resources and to determine whether or not they are having income.
- The records are essential for planning and decision making
- The records are used in obtaining credit
- The records are used for income tax

### 3.9.4 Pesticide performance records

- To show that the pesticides have been used appropriately, safely and legally, the producers must keep the records.
- It is good practice for all producers/users to keep records as it will help in good management.

- Keeping record will help in providing people with important information, in case of emergencies when people, animals or environment have accidentally been contaminated or crops have been damaged.
- To help growers in keeping track of periods when crops cannot be harvested and people or animals cannot enter in treated area.
- To meet specific conditions of crop assurance schemes or woodland assurance standard.

# Chapter 4

# Weed Control and Management in Bulb Crops

## 4.1 Identify types of weeds found in bulb crop

### 4.1.1 Weeds

- Bulb crops are grouping which includes all allium species except chives.
- Bulb crops include onions (dry and green), leeks, garlic, and shallots.
- Weed is a plant considered undesirable in a particular situation.
- It includes plants unwanted in human-controlled settings like farm fields, gardens, lawns and parks.
- Weeds can be grouped as grasses, sedges, lilies and broadleaves.
- It has no botanical significance as a plant that is a weed in one context is not a weed when growing in a situation where it is in fact wanted, and where one species of plant is a valuable crop plant, another species in the same genus might be a serious weed, such as a wild bramble growing among cultivated loganberries.
- Volunteer crops are regarded as weeds in a subsequent crop.
- Many plants that people widely regard as weeds also are intentionally grown in gardens and other cultivated settings, in which case they are sometimes called beneficial weeds.
- Weed applies to any plant that grows or reproduces aggressively, or is invasive outside its native habitat.

**Features**

- A number of weeds like dandelion and lamb's quarter, are edible with leaves or roots used for food or herbal medicine.
- It prevent pest insects from finding a crop as their presence disrupts the incidence of positive cues which pests use to locate their food.
- Weeds act as a living mulch and provides ground cover to reduces moisture loss and in preventing erosion.

- It improves soil fertility by bringing up the nutrients like calcium and nitrogen from deep in soil with tap root and clover hosts nitrogen-fixing bacteria in its roots, fertilizing the soil directly.
- The dandelion species breaks up the hardpan in overly cultivated fields and help crops to grow deeper root systems.
- Many garden flowers originates as weeds in cultivated fields and have selectively bred for garden worthy flowers or foliage.

## Characteristics

- Controlling weed development during bulb crop cycle is essential to get high yields and marketable products.
- The composition of weed population is highly variable which depends on soil type, crop rotation, sowing and growing period etc.
- Weed control is challenging in bulb crops as few herbicides are registered since mechanical cultivation is limited in high density plantings and handweeding results costly.
- Onion and garlic poses challenges to weed management as these crops are sown at high plant densities and are not thinned to produce maximum possible yield per acre.
- Onions are planted with 4 to 10 seed lines on beds that are 40-80 inches wide or 36 inches wide in intermountain area.
- Garlic is planted with 2-4 seed lines on 40-inch beds and such planting configurations limit cultivation to furrow which narrows row middles on bed top.

## Types of Weeds

- There are three types of common weed plants in regards to their growing characteristics which include:

Annual types

- It germinates and spread by seed having average lifespan of less than an year.
- It includes winter and summer types.
- It grows from seed, develop into mature plant, set flowers and seeds and die after seeds gets shed.
- They maintain their populations by production of seeds with each generation.

- ❖ Winter annuals like chickweed, germinate in late summer/early fall, go dormant in winter and actively grow during spring.
- ❖ Summer annuals like lambs quarters, germinate in spring, grow throughout summer and goes with the arrival of cold weather.

Biennial types

- ❖ It completes their life cycle in two years, germinates and forms rosettes in first year and produce flowers and seeds in second year.
- ❖ These weeds are best controlled in young growing stages of first-year plant.
- ❖ It includes bull thistle and garlic mustard.

Perennial types

- ❖ These plants are able to live for two years or more.
- ❖ It returns every year and produces long tap roots with seeds.
- ❖ It includes dandelions, plantain and purple loosestrife.
- ❖ It is most difficult to control.

### 4.1.2 Types of weeds found in bulb crop

#### Common weeds found in onion field

Purple nutsedge

- ❖ It has perennial sedge.
- ❖ Its stem is hard, fragrant, globose-ovoid tubers and is 1.2 cm long with 0.3-0.7 cm in diameter which is generally erect and 10-75 cm long.

- Its leaves are linearly narrow, shorter or longer than stem which is 0.4-0.8 cm wide with dark green above and pale beneath.
- Its flower is inflorescence simple or compound umbel with 2-7 rays each having 7.5 cm length, bearing short spikes of 3-10 spreading having red-brown spikelets.
- Its fruit has oblong to ovate-oblong shape with 1.3-1.5 mm long and 0.5-0.7 mm wide which matures to brown colour.

Common lambsquarter

- It is a polymorphous, non-aromatic and erect herb.
- Its has angular stem which is 0.3-3 m tall that is normally striped green, red or purple in colour and is clammy to touch.
- It has varied size leaves normally with upper leaves rhomboid, deltoid to lanceolate while lower leaves as toothed or irregularly lobes having 10-15 cm length, having petioles as long as leaf blades.
- Its flower are green which are borne in clusters that forms compact/loose panicled auxiliary spike.
- Its seeds are round, black and shining.

Swine-cress

- It is an annual herb having its stem repent, decumbent or ascending that radiates from central point which is glabrous, terete and green.
- Its leaves are alternate, petiolate, pinnate with 4-5 cm long and 2 cm wide having lobed, acute, linear elliptical to linear oblong shape.
- Its flower is inflorescence having small raceme with length of around 4 cm, opposite to stem leaves and are compact.

Canary grass

- It is a tufted annual bunchgrass which is around 1.8 metres high.
- Its stem is erect or decumbent and caespitose
- Its has long, linear, acuminate leaves with oblong hyaline membrane lingule and is about 2-5 mm long, truncated and/or fringed having smooth sheath.
- Its panicle more or less protruding from uppermost swollen leaf sheath, ovate to oblong with 5-8 cm long and is green.
- The flower is sikelets green which is lanceolate on short pedicels having 4 -6 mm long.

Crowfoot grass

- It is an annual, very variable, grass which is normally 10-44 cm high.
- Its stem is erect or creeping culms that roots from profusely branched nodes.
- It has linear, tapering leaves which is 2-10 cm long and 0.2-0.4 cm wide having flat, glaucous, glabrous or hispid shape with leaf sheaths striate having short ligules membranous.

- Its flower is inflorescence with 2-6 digitate spikes and 0.5-4 cm long with olive-grey colour. It has spikelets of 2-5 flowered that spreads at right angles, pendulous, strongly striate.

Scarlet pimpernel

- It is a low growing annual having height up to 30 cm.
- It has branched, decumbent, repent or erect stem that sometimes root at nodes.
- It has opposite, sessile, ovate leaves which is 2 cm long, 1.3 cm broad, having brown speckles that is variously pubescent, margins and tuberculate.
- It has solitary flowers that arise from area between stem and leaves which occur on long stalks having bright blue colour.
- It has brown seeds that are 1.3 mm long.

Smart Weed

- It is prostrate, densely branched annual herb.
- It has branched stems from base having many branches.
- Its has lance like, elliptical leaves which are 1.0-3.0 x 0.1-0.4 cm.
- The ochrea in leaves are 1-2 mm long.
- Its leaves are membranous, ovate and tube-shaped.
- The flowers are inflorescence auxiliary that sunks between ochrea with stalkless; pinkish red colour.

Bermuda grass

- It is a perennial grass having rhizome stem in top 10 cm of soil.
- The stems spread horizontally in wider distance, with nodes at 10 cm intervals, having 2-3 scale leaves with one auxiliary bud.
- In dense stands, shoots develop from buds on rhizomes or runners which tends as erect and short having 25 cm height, which develops into prostrate runners under less dense conditions.
- The leaf blades are dull grey-green in colour with flat having height upto 15 cm and width upto 3-5 mm.
- The leaves are tapered at abrupt point with fine parallel ribbed surfaces.
- Its flower is inflorescence that supports on culm to 25 cm height having single whorl of 3-7

## 4.2 Maintain records of the weed and share it with experts

### 4.2.1 Record Keeping

The ability to compare and reflect on data from worst crop to best crop is vital. Bulb crop grower must always record, every aspect of their cultivation facility, whether indoor, outdoor, greenhouse, organic or anything in between. Having the ability to compare data from past harvests is priceless as it allows crop grower to analyze the nuances and differences between varying results and attempt to explain any inconsistency, so crop grower should record every possible aspect of their facility, even if it's just hand-written on paper and should always discuss with their superiors for betterment.

### 4.2.2 Necessity of maintaining records

The crop grower/producers should keep the following weed records for:

- Nutrient management plans
- Crops planted and actual yield
- Timing, location and crop acreage of all nutrient applications
- Analysis of nutrient content of any fertilizer applied
- Receipts related to purchase of nutrients
- Animal waste generation measurements and estimations
- Justifying any changes from nutrient management plan as written

### 4.2.3 Items to be recorded and discussed

Bulb crop grower should record following information on data sheet which covers:

- Installation Date of Light Bulbs: Bulb crop grower must determine the point of diminishing returns for all lightbulbs, that gives life expectancy.
- Light Intensity: Light intensity levels must also be recorded as part of work schedule with which, bulb crop grower can decide exactly when to replace lightbulbs to ensure maximum intensity and minimize potential yield losses.
- Periodic Water Analysis: Bulb crop grower should analyse water source at least two times in an year and should record.
- Write To-Do List: Bulb crop grower should make goals and plan by asking themselves, how they will achieve the goal and plan their strategy by writing with target date to do so.
- Cost of Goods: Bulb crop grower should record every possible cost of cultivation and business which allows them to analyse exactly where they are spending their money. He/she can then make decisions that are more conducive to minimize unnecessary expenditure and/or waste.
- List of inventory and tracking materials: It's much more efficient to know exactly how much of given item is on hand and to order more before it's needed, rather than be forced at the last minute to source whatever is necessary for day-to-day operations. Bulb crop grower should record the inventory and make purchasing department aware when inventory of any item is close to depletion.

- Workstation Lists: Bulb crop grower/managers should have lists of all materials required to properly maintain and fulfill their responsibilities. They should be required to submit their list weekly to purchase department to ensure seamless workflow.
- Power Usage: By determining exactly when bulb crop grower are using most expensive rates, as they may be able to slightly alter some operations or aspects to save capital. They may choose to use battery storage system which allows them to offset energy costs by charging during off-peak hours, so bulb crop grower can use stored energy during peak-demand hours.
- Water Temperature: Bulb crop grower should record the temperature of water given to any plant as neither too cold nor too hot water is best, as in both cases, water holds less oxygen. At 62°F, water holds maximum oxygen, but at 62°F, it is cooler as preferred by plants, but very hot temperatures are not preferred.
- Potential of Hydrogen (pH): Bulb crop grower should keep accurate records of water before and after any nutrient or amendment is applied and in case of any adjustment of pH up or down, to reach desired level. As water source changes, adjustments need change.
- Water Parts Per Million (ppm): Bulb crop grower should record ppm of water before and after any nutrient or amendment application for all stages of growth.
- Nutrient Log: Bulb crop grower should write every aspect of nutrient application, ie how much, of what, day it was applied, etc as it goes for every stage of growth for every amendment, whether it be organic or salt-based, indoor, outdoor, greenhouse or even compost teas and foliar feeds.
- Daily Activity Log: Bulb crop grower should record daily activities related to crop growth, from work performed, to number of hours it took to do each task. With such data, bulb crop grower can see exactly where the most time is being spent and all associated costs of performing a specific task.
- Air Temperatures: Bulb crop grower should record highs and lows of outside ambient temperatures, and highs and lows in any given 24 hours for all cultivation environments. Monitoring and recording nighttime environment is important as monitoring daytime environment, so bulb crop grower must always understand what is taking place at night so that they can adjust accordingly.
- Humidity: Bulb crop grower should log the humidity highs and lows of outside environment and highs and lows of any given 24 hours for all cultivation environments as recording of daytime and nighttime highs and lows will help in monitoring the nighttime environment and daytime levels. Since plants transpire and release moisture at night and in day, so humidity levels may become elevated and promote conditions conducive to fungal or mildew growth and proliferation.
- Air Content: Bulb crop grower should record and log all possible air-quality levels. CO2 and oxygen levels must be constantly monitored and recorded. Ethylene gas can build up, and be detrimental and counterproductive in many stages of plant growth. Bulb crop grower must have proper airflow and accurately monitor air quality as proper airflow will aid in air-moisture-buildup prevention.

- Dissolved Oxygen Levels of Water and Oxygen Content: Bulb crop grower should record and constantly monitor oxygen levels in all water that will be used on plants. Factors like water temperature will affect the DO levels in water.
- Yield Per Square Foot of Plant Canopy: The only way to know if the facility is improving or not is to record the amount of usable cannabis produced in each area. Each individual room, plot or greenhouse is a separate entity with its own unique nuances, so all rooms and environments are not exactly same. A small degree of variation always exists and each environment is separate and should be recorded and logged as such.

Bulb crop grower should accumulate and record all necessary data and value so as to made any refinements. The data should be automatically recorded on a computer platform that would allow bulb crop grower to compare the best harvest to worst harvest, which gives a true perspective on why one crop did better than another. Recording of data is also useful in case of any problem or when problem arises. The goal is to give the plant exactly what it wants exactly when it wants it.

## 4.3 List them as per broad leaf and narrow leaf weed

Weeds interference in vegetable crops intensifies due to cultivation areas that go through intensive exploitation, frequent tillage, high fertilization rates and low water restriction. Such environment is favorable for occurrence of weeds, which are fast-growing, having shorter development cycle and high production of diasporas that significantly increase soil seed bank.

| Broad leaf weed | Narrow leaf weed |
|---|---|
| Native rosella | Bristly starbur |
| Musk mallow | Desert cotton |
| Florida keys Indian mallow | Mountain knot-grass |
| Country mallow | Indian jointvetch |
| Indian copperleaf | Flossflower |
| Devil's horsewhip | Red calico plant |
| Goat weed | Smooth joyweed |
| Brazilian joyweed | Alligator weed |
| Khaki weed | Sessile joyweed |
| Sweet alyce clover | Long leaved alyce clover |
| Spiny pigweed | Alyce clover |
| Slender amaranth | Over-leafed alysicarpus |
| Scarlet pimpernel | White moneywort |
| Coral vine | Blistering ammannia |
| Ganges primrose | Kalmegh |
| Showy pigeonpea | Indian catmint |
| Crested Philippine violet | Malabar catmint |
| Porcupine flower | Mauritian grass |

| Broad leaf weed | Narrow leaf weed |
|---|---|
| Para cress flower | Mexican prickly poppy |
| Spiny leaved blumea | Giant reed |
| Red spiderling | Onion weed |
| Shaggy buttonweed | Waterleaf |
| Browntop millet | Wild oat |
| Crown flower | Blanket grass |
| Swallo-wort | Jerry water-fire |
| Hemp | Cape ash |
| Pig's senna | Spanish needles |
| Coffee senna | Creeping Blepharis |
| Dwarf cassia | Soft blumea |
| Sicklepod | Erect spiderling |
| Bush Grape | Tiny false buttonweed |
| Asiatic pennywort | Annual brachiaria |
| Butterfly pea | Sweet signal grass |
| Common lambsquarter | Running grass |
| Nettle-leaved goosefoot | Pink node flower |
| Siam weed | Hairy bittercress |
| Giradol | Balloonvine |
| Rottler's chrozophora | Feather-leaved cassia |
| Tick weed | Tiny periwinkle |

## 4.4 Control weeds physically or chemically

### 4.4.1 Preventative Weed Control

❖ It refers to control method that aims to prevent weeds from being established in a cultivated crop, a pasture, or a greenhouse.

❖ Examples include use of certified weed free seed, transporting weed free hay, making sure farm equipment is cleaned before moving from one location to another, and screening irrigation water to prevent weed seeds from traveling along irrigation ditches.

Preventative practices of weed include:

❖ cleaning tillage and harvesting equipment of weed seeds and vegetative structures

❖ planting certified

❖ weed-free crop seed

❖ controlling weeds in barnyards, around structures and along fencerows, roadways and ditch banks

### 4.4.2 Techniques of Weed Control

Controlling weeds in crop production involves wide range of techniques which includes:

**Cultural**

- It refers to a technique that involves maintaining field conditions such that weeds are less likely to become established and/or increase in number.
- Examples of cultural weed control would be crop rotation, avoiding overgrazing of pastures or rangeland, using well-adapted competitive forage species, and maintaining good soil fertility.

**Mechanical**

- It refers to technique that involves the use of farm equipment to control weeds.
- Mechanical/physical techniques destroy weeds or make environment less favorable for seed germination and weed survival.
- Such techniques include hand-pulling, hoeing, mowing, plowing, disking, cultivating and digging.
- Mulching can also be considered as mechanical control means as it uses physical barrier to block light and impede weed growth.

**Biological**

- It refers to technique that involves the use of natural enemies of weed plants to control the germination of weed seeds or the spread of established plants.
- It is rapidly expanding area of weed control with many examples which includes sheep to control tansy ragwort or leafy spurge, cinnabar moth and the tansy flea beetle to control tansy ragwort, the chrysolira beetle to control St. John's Wort, and the use of goats to control brush on rangeland.

**Chemical**

- It refers to technique that involves the application of a chemical (herbicide) to weeds or soil to control the germination or growth of the weed species.
- It includes chemicals to control weeds which includes 2,4-DB; EPTC; bromoxynil; and paraquat.
- Herbicides are crop protecting chemicals used to kill weedy plants or interrupt normal plant growth.
- Herbicides provide convenient, economical and effective way to help manage weeds and allow fields to be planted with less tillage showing earlier planting dates and provide extra time to do other tasks that farm/personal life require.
- With reduced tillage, soil erosion reduces which reduces soil from entering waterways and decreases quality of surface water.
- Without herbicide use, no-till agriculture becomes impossible, but use of herbicide also carries environmental, ecological and human health risks.

- Herbicides may not be necessity on some farms/landscape settings, but without use of chemical weed control, mechanical and cultural control methods become more important.
- There are many kinds of herbicides that can be selected based on many factors that determine when, where and how particular herbicide can be used most effectively.

## 4.5 Undertake weeding process at appropriate time to avoid crop damage

- Weeds compete with crops for space, nutrients, water and light and causes cash crop to produce less yield.
- To produce high yields, weeds should not be allowed to compete with the crop as presence of weeds does not necessarily damage a crop, especially during early growth stages when both weed species and crops grow without interference.
- Killing of weeds early in their life cycle and not allowing them to produce seeds, dramatically reduces weed pressure.
- Weeds, along with quality seed, timely planting, soil moisture, and soil fertility, have the largest impact on crop yields.
- Timely weed removal early in growing season improves production, so in limited rainfall, it is important, not to allow weeds to compete with more valuable crops.
- The best time to weed is when the soil is moist and the weeds are very small, mere seedlings.

## 4.6 Use herbicides for weed control wherever feasible

### 4.6.1 Chemicals/herbicides

- Herbicides play a vital role in integrated weed management programs.
- Knowledge of the mechanisms and activity of herbicides will improve the impact and sustainability of herbicides as a weed management tactic.

### 4.6.2 Types of herbicides

**Translocated herbicides:**

- They move to the site of action by transport mechanisms in plant; xylem and phloem.
- The xylem transports water and nutrients from the soil to growth sites and phloem transports products of photosynthesis (for instance, sugars) to growth and storage sites.
- It may take up to two weeks for symptoms to develop on the target weeds depending on herbicide rate, conditions and species.

**Contact herbicides**

- They have limited movement within the plant, so complete coverage of target is critical.
- Compared to translocated herbicides (glyphosate), contact herbicides (paraquat, oxyfluorfen, diquat and bromoxynil) tend to show symptoms rapidly, usually within 24 hours.

**Selective herbicides**

- ❖ It kill target weeds and not desired plants (crop or pasture) when applied at specified application rate.

**Non-selective herbicides**

- ❖ It is also called as knockdown herbicides like glyphosate or paraquat that damage most plants.

**Residual herbicides**

- ❖ It remains active in the soil for an extended period of time and acts on successive weed germinations.

**Non-residual herbicides**

- ❖ It includes non-selective paraquat and glyphosate.
- ❖ It has little or no soil activity and are quickly deactivated in the soil.
- ❖ They are broken down or bound to soil particles and is less available to growing plants.
- ❖ They have little or no ability to get absorbed by roots.

**Post-emergent and pre-emergent**

- ❖ These are target and timing of herbicide application.
- ❖ Post-emergent is foliar application of the herbicide after the target weeds have emerged from the soil.
- ❖ Pre-emergent is the application of herbicide to the soil before weeds have emerged.
- ❖ Herbicide mixtures and sequential applications
- ❖ It involves application of more than one herbicide to increase spectrum of weed species controlled for resistance management.
- ❖ A mixture involves the application of multiple products in a single application.
- ❖ Where herbicides are antagonistic and cannot be mixed together in a single tank, they are applied sequentially.

### 4.6.3 Chemicals/herbicides in controlling weeds

- ❖ Chemicals that are used to kill plants or weeds are called herbicides.
- ❖ Herbicide selectivity and its dose, time and method of application are of paramount importance/consideration before applying to a crop.
- ❖ In chemical weed control, chemicals called herbicides are used to kill certain plants or inhibit their growth.
- ❖ Chemical weed control is an option in integrated weed management that refers to the integrated use of cultural, manual, mechanical and/or chemical control methods.

**Control of Weeds**

- ❖ Prevent yield loss due to weed competition
- ❖ Maintain purity and/or quality and market price of harvested grain

- Prevent build-up of weed seeds in soil
- Prevent weeds that may attract insects or rodents (rats) or act as a host for diseases
- Prevent clogging of field irrigation channels and facilitate water flow
- Reduce time and cost of land preparation and weeding operations

**Use of Herbicides to Control Weeds**

Herbicides are used to control the weeds due to:

- Less labor (0.5 person-day per ha per application) and less drudgery
- Cost-effective, if practiced properly
- No need to wait for weeds to grow bigger for hand weeding
- Herbicides can differentiate between rice and weeds, even at seedling stage where it is very difficult for people to see the difference

### 4.6.4 Control of weeds by bioherbicides

- Bioherbicide is biologically control agent for weeds which is a compounds of microbes.
- Bio-herbicides and bio-pesticides are naturals, as when active ingredient is used as fungus it results as mycoherbicide.
- Bioherbicides uses naturally occurring enemies, instead of depending on man-made chemicals as agents of biological control have many fewer, milder, effects on environment.
- The bioherbicides need not be subjected to careful tests for environmental and public health safety.
- Bioherbicides can be applied in form of aerial sprays by cut and paste application or in powder applied to soil.
- The pathogens exploited as mycoherbicides are native to area where they are used and does not need special imported.
- Under natural conditions, disease epidemics occur which damage plants from time to time, where potential of fungi is limited.
- The approach where fungi are turned into mycoherbicides, allows people to overcome constraints and create disease epidemics when and where required.
- After application, fungi do not persist at high levels for long and have often returned to background levels after 1–2 years, as like other herbicides used to kill plants.
- The pathogens used in control need not be highly host specific as classical biological control agents as their use can be restricted to certain areas.

## 4.7 Use new & innovative methods of soil solarization & mulching

### 4.7.1 Soil Solarization

- Soil solarization traps the sun's heat beneath a layer of clear plastic and disinfests soil with higher temperatures to kill weed seeds, vegetative structures, insects and disease that causes pathogens.

- Solarization controls the seedlings of bermudagrass, johnsongrass and field bindweed, and is effective on annual weeds as compared to perennial weeds.
- It partially controls yellow nutsedge, but does not control purple nutsedge.
- Annual weeds with harder seed coat like little mallow and velvetleaf are hard to kill with solarization.
- Bulb crop grower should plan to solarize soil as per planting time where plastic should remain in place for minimum 4 weeks, and for best results, should plant crop immediately after removing the plastic.
- For many crops, growers burn holes in plastic mulch and transplant directly into them, but such practice can be of limited value for high-density plantings of onion and garlic.
- As solarization needs summer fallow, it works best with fall-planted crop. In desert and Valley, plastic should be in place during June through August, and should remain in place till planting begins in fall.
- Solarization is not practical in areas with short growing seasons or in coastal areas where soil is not heated as deep. In coastal areas, use of plastic in fall when there are less chances of fog.
- The solarized soil should be cultivated less than 3 inches deep to avoid bringing viable weed seeds to surface, where they germinate.

### 4.7.2 Mulch

- Mulch helps to preserve soil moisture and inhibits the emergence of many annual weed species.
- The method can be practised by using various plant residues; some of them can be toxic to weeds.
- In this, black or transparent polyethylene sheet is kept on crop where rows are widely used.
- Before its introduction into agricultural practice, technical and economic effectiveness of plastic mulch should be evaluated in field tests.
- The use of plastic films in large areas shows the use of machinery for deploying films and disposing them after use in the crop to prevent environmental pollution.

## 4.8 Use mechanical weed equipment

**1. Rotary Hoe**

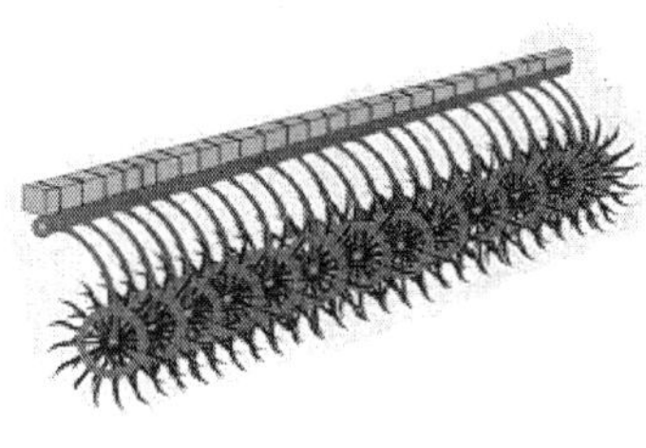

- It has dozens of closely adjacent wheels which works by throwing the soil as it rolls shallowly over the field.
- In this, many models are for high residue conditions like those found under minimum tillage systems.
- It has mild impact on the soil surface.
- It is important for use when the weeds are just coming through the soil or starting to emerge.
- In this, last weed stage lasts 4 to 8 days as per the climate.
- Here, stage of the crop is not as critical as for the finger weeder as the rotary hoe has more gentle action.
- It gives best results in slightly dry and crusted soils and also in sandy soil if it is little moist.

## 2. Finger Weeder

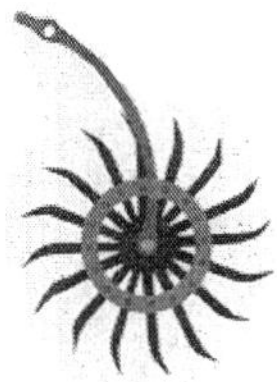

- It has long tines that vibrate in the soil.
- It is made of several sections that allow implement to follow the surface of the field.
- It has a lever that allows adjustment of all tines in a section at once.
- Depth wheels allow better precision when adjusting.
- In this, crop stage is critical due to its stronger action.
- The finger weeder is not good if lot of residue is present on the soil surface.
- It will drag the residue and damage the crop.
- Its effectiveness is reduced after first leaf-stage of weeds.

## 3. Tine weeders

- It has multiple rows of flexible metal tines that cover the entire soil surface.
- It wiggle slightly when pulled along, uprooting or dislodging small weeds.
- It works good at relatively high speeds.
- Its wide span makes the cultivation quick.
- It is rear-mounted and is available in many widths.
- In this, the tines tension can be adjusted on certain units or by pressure on 3-point hitch on other units.
- The gauge wheels are used to maintain uniform depth.

- It works best in friable soils which are free of rocks and also when weather allows weeds to dry out on soil surface after uprooting.

**4. Basket weeders**

- It is of relatively high-speed, between-row cultivators.
- It is good for control of small weeds in narrow rows when crop is small.
- The wire baskets work the soil surface and do not move soil into the row.
- It will not work well in crusted or rocky soil.
- The front set of baskets are ground driven where rear set of baskets are turned faster by chain and gearing, causing scuffing of the soil.
- It is available in 2 to 6 row units which are rear or belly-mounted.

**5. Row crop cultivator**

- It has gangs of S-shanks with sweeps that break up the soil by destroying small to moderate-sized weeds.
- The gangs work behind the tractor between the rows at uniform intervals.
- In this, shields are added to protect the crop from moving soil.
- In this, fertilizer hopper is used for side dressing during cultivation.

**6. Rolling cultivator**

- It has gangs of soil driven spider wheels which is kept independently on toolbar.
- It has an angle which works the soil with aggressiveness that can be adjusted.
- In this, numbers of gangs are grouped together which shows cultivator widths that are rear-mounted.
- In this, pairs of gangs are belly mounted to work a row or two.
- It throws the soil in row to bury small weeds or to form hills as per the angle of gangs.
- It is a relative heavy and aggressive tool.

**7. Wiggle hoe**

- It is in-row cultivator which uses hand steered side knives that is controlled by operator who moves it, in and out of the row.
- In this, the knives reach close to the row and remove the weeds, but when operator is careless, the crop can get damaged.
- It is good for irregular plant spacing in the row or between the rows.

**8. Reigi weeder**

- It is a new tool that needs a rear operator who steers a pair of rotating horizontal wheels in and out of crop row.
- In this, wheels are turned by PTO-driven belt having stiff tines that root out the weeds.

- It has wheels which come in many sizes for different row spacing's while the units come in 1 or 2-row models.
- It is good for killing weeds in and next to the row in widely spaced crops.

## 4.9 Optimum stages of weed control i.e. pre emergence or post emergence

### 4.9.1 Pre emergence weed control

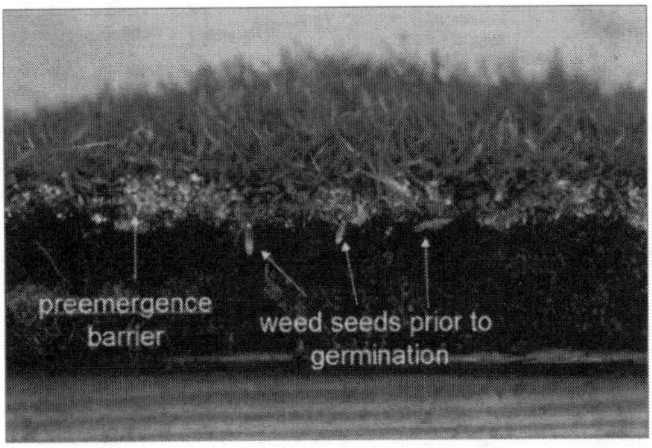


**Introduction**

- In India, main weeds in winter grain crops are Board leaved weeds such as Lamba quarters (Chenopodium sp.) and Wild safflower (Carthomus Osycantha)
- In some parts of the country, perennial dicot weeds such as canda thistle (Cirsium avenses), field bind weed (Convalvelus arvensis) are found.
- Annual Grasses like Wild Oat (Avenu fatua), Chanry grass (Philaris minor).
- The weed seeds get admixture in the crop seed and reduce grain quality.

**Pre-emergence weed killers**

- These are used on established fields as part of annual effort to combat plant pests.
- These chemical compositions are used before weeds take hold to kill off infant root systems and keep them from growing.
- It is used before seeing weeds to save them from showing up in fields which has no interference of chemicals with germination but can save formation of new root cells in baby weed plants.
- Timing, weather and the type of weeds that are problematic in fields will dictate the exact formula and application for using pre-emergence procedures.
- Chemicals used in pre-emergence are not effective on vegetative buds that sprout from existing roots or rhizomes.
- The chemicals cannot be used on prepared grass seed bed as their root stunting action in young plants also affects sprouting grass.
- Grown crop has no fear, as their root system is already developed and the crop is hearty and healthy.

- Pre-emergent procedure shows that it is the sensitive root tissue of newly germinated seedlings which is killed off, that result in complete plant death.
- Perennial weeds develop thick persistent adult roots that re-sprout in spring, which makes them difficult to control with pre-emergent formula.
- Timing of pre-emergence weed killer should match with germination period for the variety of weed.
- Biennial weeds are not controlled by pre-emergence procedure as they produce seed that germinates nearly year around.
- These weed killers need water to activate themselves and carry the chemical down to the root systems of newly sprouted weeds.

### 4.9.2 Post emergence weed control

#### Introduction

- Post-emergent's attack the weeds once they grow up.
- Post-emergence herbicide controls the existing weeds and helps in preventing future ones.
- There are different types of post-emergent weed killers, what cultivator should keep information related to variety of weeds and need to control them.
- Post-emergent weed killers attack the foliage or flow systemically to the roots of the weed.
- Herbicide application after the emergence of crop or weed is referred as post-emergence application.
- When the weeds grow before the crop plants have emerged through the soil and are killed with a herbicide then it is called as early post-emergence.
- Spraying 2,4-D Na salt to control parasitic weed striga is called as post-emergence while spraying of paraquat to control emerged weeds after 10-15 days after plantation can be called as early post-emergence.
- Eg. Glyphosate, Paraquat, 2,4-D Na Salt.

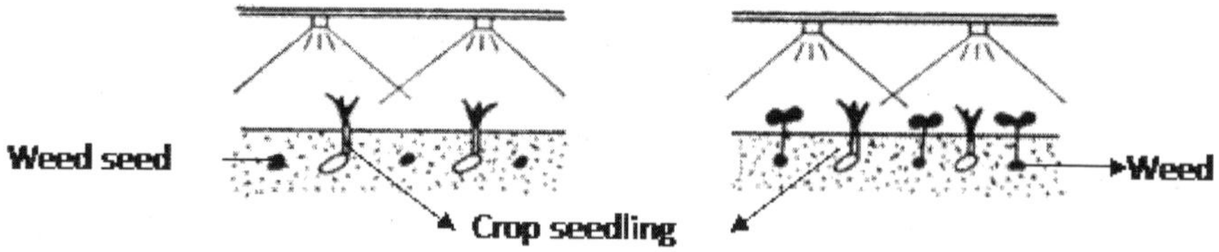


#### Types of Post-Emergent Weed Killers

- The post-emergent can be either systemic or contact applications.
- Systemics are useful on perennial weeds as they are absorbed directly in plant and move throughout it for maximum killing action.
- Contact herbicides kill the exposed part of plant and are used on annuals and smaller weeds which kill the entire plant.

## Post emergence herbicides

- These herbicides are applied directly to emerge weeds.
- In contrast to pre-emergence herbicides, post emergency herbicides controls only weeds that are emerged and actively growing at the time of treatment.
- They are used for farm weed control which is systematic or translocated in the plant system that transports food and water.
- It is applied after the weeds and crop have emerged from the soil.
- The treatments can be applied in either a broadcast or directed fashion.
- When applying post emergence herbicides, it is necessary to have maximum coverage of the weed with the spray solution.
- Surfactants are used with post emergence herbicides to enhance control.
- These herbicides are classified as selective and non-selective.
- Selective herbicides are used to target certain weeds, and in areas such as turf where contact with the grass is unavoidable.
- Non-selective herbicides are used for broad weed control and have a purpose in open, unmanaged fields, for instance.
- They control weeds that have emerged since crop or pasture establishment and can be applied with little damage to the crop or pasture plants.

### Benefits

- It gives high level of target weed control with extra benefit of improved crop or pasture yield.
- It allows fine-tuning of herbicide selection to match weeds present in the paddock.
- In this, timing of application is flexible to suit weed size, crop growth stage and environmental conditions.
- It has pre-emergent activity on subsequent weed germinations.

### Issues

- It requires careful consideration when selecting the best post-emergent herbicide in any one situation.
- Use of post-emergent herbicides to stressed crops and weeds results in reduced levels of weed control and increased crop damage.
- In this, crop competition is important for effective weed control by selective post-emergent herbicides.
- It requires suitable technique for the situation to optimise control.
- It requires correct adjuvant to ensure effective weed control.
- It needs suitable formulation of herbicide for every situation.
- The effectiveness of post-emergent herbicides is influenced by range of plant and environmental factors.

## Chapter 5

# Irrigation Management in Bulb Vegetable Crops

### 5.1 Requirement of pre sowing irrigation

- The pre-sowing irrigation resulted in stronger seedlings and larger and deeper root systems, compared with non-irrigated treatments.
- Pre-sowing irrigation significantly increases water productivity of spring crop due to substantial improvement in soil moisture at deeper soil profiles.
- Pre-sowing irrigation application results as effective practice for agricultural water management in semi-arid areas.
- It is commonly used by local crop growers to promote germination and crop establishment due to low and uneven seasonal rainfall.
- In case of dwarf high yielding varieties, pre-sowing irrigation should be given and crop-sown when the field becomes fit for operation.

### 5.2 Irrigation after crop transplanting

#### 5.2.1 Transplanting

- Transplanting is a method of planting seeds in seed box which on bigger seedlings gets transplanted from seed box to garden plot.

- It is done when seedlings attains 12 -15 cm height with development of 3 – 4 leaves.
- Transplanting is planting with the use of pre-grown seedlings or plants that had been propagated from seeds.
- Crop area is the space which is particularly used for growing crops.
- It is indirect as seeds are not immediately sown, but are first used to raise the seedlings in pots or in temporary place which are further used when reaches at right age.
- Field planting is done either manually or by mechanical means.

### 5.2.2 Irrigation on crop

- In initial month of May, seed get transplanted in field beds at a distance of 20 cm x 10 cm and attains the height of 12 to 15 cm, so field should be irrigated just after transplanting.
- For better plant growth and good bulb formation, sufficient watering is done in 7-10 days interval.
- The irrigation should be stopped 20 days before harvesting for proper maturity of bulbs.
- It is essential to protect bulbs from low temperature during night at time of harvesting as it causes spoilage of bulbs in field.

## 5.3 Irrigation as per climatic condition

- Climate change not only affects water availability but also the demand for water.
- If climate in given region gets drier and warmer, water availability will decrease and be exacerbated by increasing water demand.
- Soil water content and therewith irrigation in agriculture is likely to be affected by higher temperature.
- Climatic parameters changes will affect the demand for water and supply.
- Changes in water demand, irrigation practice enhance groundwater exploitation which shows that climate change to impact the groundwater resources.
- As temperature increases, the atmospheric air gets warmer which accelerates the evaporation from soil moisture.
- Irrigation requirements are determined by climate and by crop and soil characteristics and is applied when soil water content is lower than 20% of available water capacity.
- There is no need of irrigation in high-rainfall as precipitation water is sufficient to grow the crop.

| Climatic Factor | Crop water need | |
|---|---|---|
| | High | Low |
| Sunshine | sunny (no clouds) | cloudy (no sun) |
| Temperature | hot | cool |
| Humidity | low (dry) | high (humid) |
| Windspeed | windy | little wind |

- The highest crop water needs are found in areas which are hot, dry, windy and sunny.
- The lowest values are found when it is cool, humid and cloudy with little or no wind.
- Single crop grown in different climatic zones will have different water needs.
- For calculating water need of crop, it is useful to take a certain standard crop or reference crop and determine how much water is required by the crop per day in various climatic regions.

## 5.4 Irrigation systems/methods

The irrigation system to be used will be determined by availability of water, topographical and soil conditions. When initial soil preparation is complete, the installation of required irrigation system will be implemented as per prescribed design.

The irrigation system consists of:

**Intake structure**

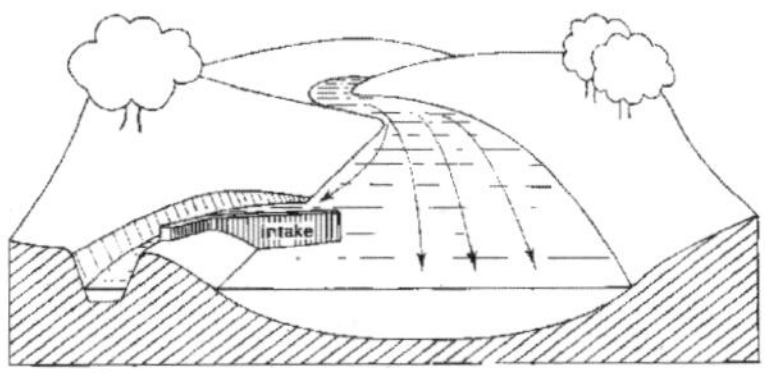


- The intake structure is built at the entry to the irrigation system.
- The purpose of intake structure is to direct water from the original source of supply (lake, river, reservoir etc.) into the irrigation system.

**Pumping station**

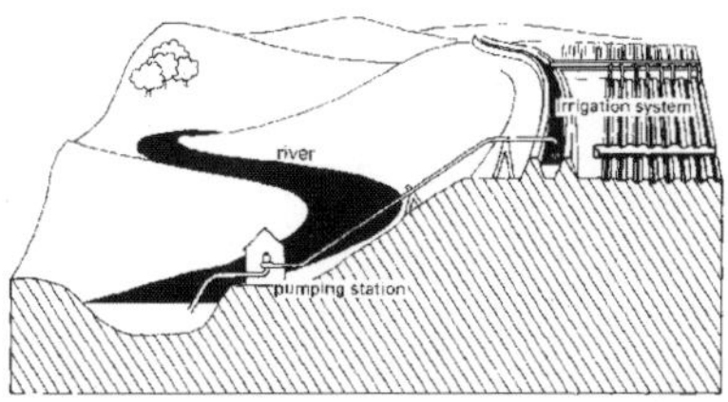


- Many times, irrigation water source lies below the level of the irrigated fields.
- The pump is used to supply water to the irrigation system.
- There are several types of pumps, but the most commonly used in irrigation is the centrifugal pump.

Conveyance system and distribution system

- The conveyance and distribution systems consist of canals transporting the water through the whole irrigation system.
- Canal structures are required for the control and measurement of the water flow.

Open canals

- An open canal, channel, or ditch, is an open waterway whose purpose is to carry water from one place to another.
- Channels and canals refer to main waterways supplying water to one or more farms.
- Field ditches have smaller dimensions and convey water from the farm entrance to the irrigated fields.
- The most commonly used canal cross-section in irrigation and drainage, is the trapezoidal cross-section.

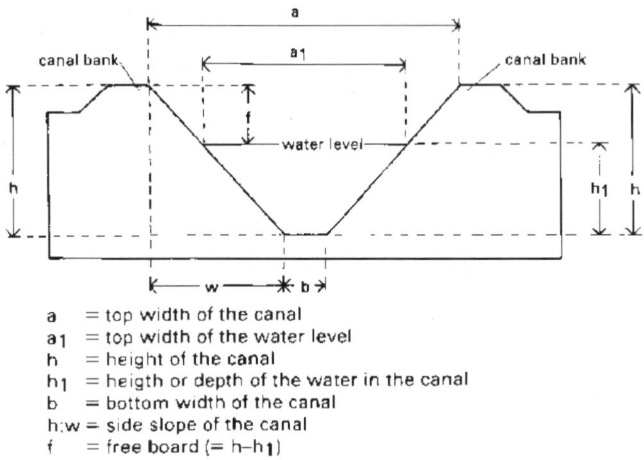


Field application system

- ❖ The traditional methods of applying water to the field is by bringing the water from the source of supply, such as a well, to each plant with a bucket or a water-can.
- ❖ Surface irrigation is application of water to the fields at ground level in which, either the entire field is flooded or the water is directed into furrows or borders.
- ❖ Furrows are narrow ditches dug on the field between the rows of crops.
- ❖ The water runs along the furrow as it moves down the slope of the field.
- ❖ The water flows from the field ditch into the furrows by opening up the bank or dyke of the ditch or by means of syphons or spiles.

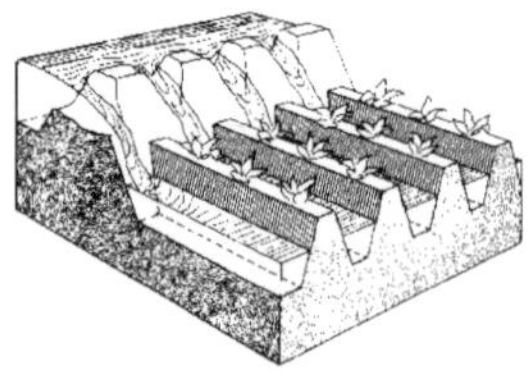

- ❖ Siphons are small curved pipes that deliver water over the ditch bank.

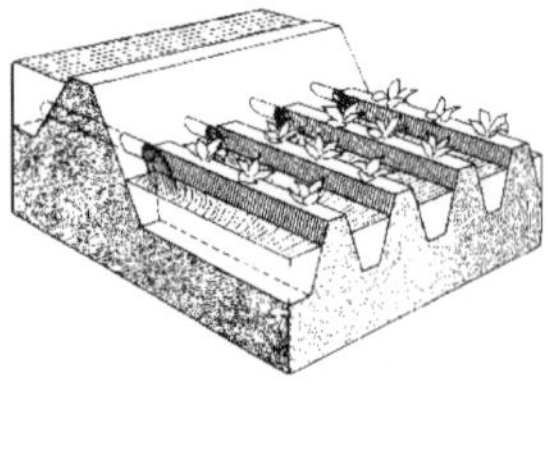

- Spiles are small pipes buried in the ditch bank.
- In border irrigation, the field to be irrigated is divided into strips by parallel dykes or border ridges.

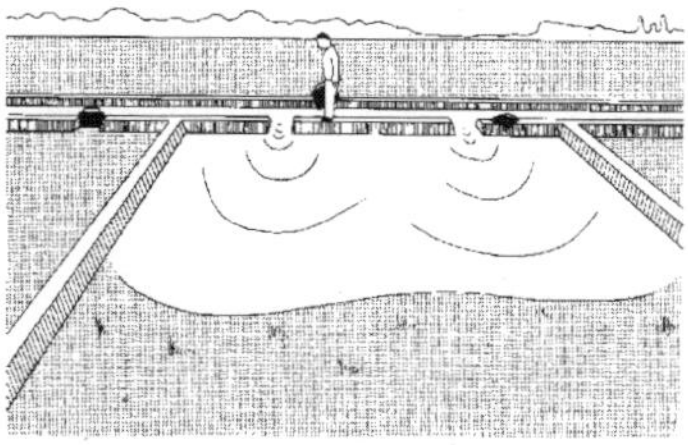

- In border irrigation, water is released from the field ditch on border through gate structures called outlets or by means of siphons or spiles.
- In basin irrigation, basins are horizontal, flat plots of land, surrounded by small dykes or bunds in which the banks prevent the water from flowing to the surrounding fields.

- Basin irrigation is commonly used for rice grown on flat lands or in terraces on hillsides.

- Trees can also be grown in basins, where one tree usually is located in the centre of a small basin.

Drainage system

- A drainage system is required to remove excess water from the irrigated land.
- The excess water may be e.g. waste water from irrigation or surface runoff from rainfall.
- Waste water can also be leakage or seepage water from the distribution system.
- Excess surface water is removed through shallow open drains.
- Excess groundwater is removed through deep open drains or underground pipes

## 5.5 Irrigation as per soil characteristic

### 5.5.1 Irrigation system based on soil type

- It is seen that water is absorbed and moves slowly through clay soils, but once the soil gets wet, they retain significant amounts of moisture.
- Water is absorbed and moves quickly through sandy soils, but they retain very little which shows that water is applied quickly to clay soil has the tendency to run off rather than move in the soil.
- Irrigating of clay soils shows that water needs to be applied slowly over long period as a result of which, the site will not need irrigation for several days.
- Irrigation on sandy soils is applied quickly but for short periods.
- Irrigation times on sandy sites should be shorter, otherwise water moves beyond the root zone, and becomes unavailable to the plant and contributing to soil leaching.
- For efficient water use under certain weather conditions, sandy sites may need daily irrigation for short periods.
- Clay soils have greater capillary movement as compared to sandy soils.
- Quick water application on sandy soils will contribute to big wetting area which gives more soil volume for roots to exploit.

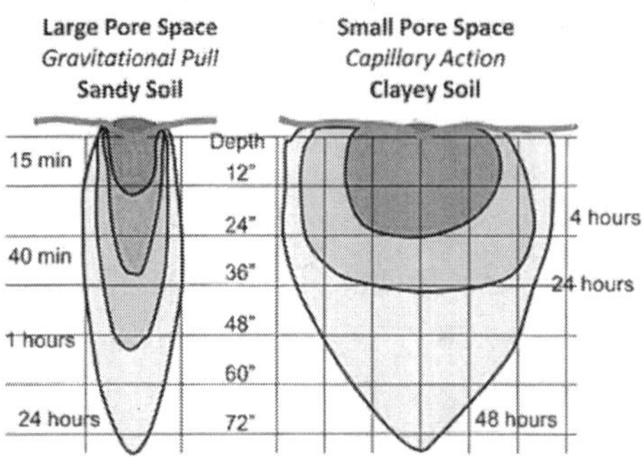


- Clay-based soils have small, flat, compact particles with large surface to volume ratios.
- Clay soil are often difficult to prepare for planting since they are slippery when wet and hard when dry, making timing for field operations critical to avoid damaging soil structure and getting proper soil tilth for planting.
- Sand-based soils are at the other end of the spectrum having comparatively large particles with small surface to volume ratios.
- Sand-based soil are easier to prepare for planting and can be worked shortly after significant rainfall.

### 5.5.2 Irrigation Technology as per soil

Basin irrigation

- It is the simplest of the surface irrigation methods.
- If the basins are small, they can be constructed by hand or animal traction.

- Their operation and maintenance is simple.

Furrow irrigation

- It requires accurate field grading which is mostly done by machines.
- The maintenance such as ploughing and furrowing are done by machines.
- It needs skill, organization and frequent use of foreign currency for fuel, equipment and spare parts.

Short level furrows

- It is called as furrow basins.
- It is like basins which can be constructed and maintained by hand.

Borders

- It requires highest level of sophistication.
- They are constructed and maintained by machines.
- The grading needs to be accurate.
- Machine operation requires a high level of skill, organization and usually foreign currency.

| Soil type | Rooting depth of the crop | Net irrigation depth per application (mm) | Irrigation method |
|---|---|---|---|
| Sand | shallow | 20-30 | short furrows |
| | medium | 30-40 | medium furrows, short borders |
| | deep | 40-50 | long furrows, medium borders, small basins |
| Loam | shallow | 30-40 | medium furrows, short borders |
| | medium | 40-50 | long furrows, medium borders, small basins |
| | deep | 50-60 | long borders, medium basins |
| Clay | shallow | 40-50 | long furrows, medium borders, small basins |
| | medium | 50-60 | long borders, medium basins |
| | deep | 60-70 | large basins |

## 5.6 Depth of irrigation and critical stages of crop growth for irrigation

### 5.6.1 Irrigation Depth

- The irrigation depth is the amount of water that needs to be applied to an irrigated system when soil water is reduced to the specified depletion level which is shown in gallons for every 1000 square feet.
- If more water is applied than is needed to bring the moisture level upto field capacity the additional amount is lost by deep percolation.
- The net amount of water to be applied at every irrigation is amount which needs to be applied to surface to be sure enough water to enters which is held in soil to meet net requirement for each irrigation.

### 5.6.2 Critical stages of crop growth for irrigations

- Most favorable soil moisture for plant growth varies with the stage of crop growth.
- Certain periods for crop growth and development are most sensitive to soil moisture stress as compared with other.
- The period is known as moisture sensitive periods.
- Critical period is used to show stages of growth when plants are most sensitive to shortage of water.
- Inadequate water supply during moisture sensitive periods will irrevocably reduce the yield and provision of adequate water and other management practices at other growth stages will not help in recovering the yield lost.
- In vegetables, transpire less water requires stress free moisture condition due to weak root system that is sparsely distributed and located in 15 to 20 cm layer of upper soil which gets dried up fast.
- Vegetable crops uses and transpire more water in later stages of growth during which moisture stress markedly reduces yield.
- All stages of growth is sensitive to soil moisture stress for crops where vegetative parts are of economic importance.
- Total growth and yield of perennial plants are sum effects of stress at each growth stage.
- Adequate water supply is required at flower bud initiation, flowering and fruit set.
- Flower bud formation, increases due to restricted water supply prior to flower bud initiation.
- The maximum benefit from scarce irrigation water is realized when irrigation is to be scheduled at moisture sensitive periods by withhold irrigations at other periods of lesser sensitivity.

## 5.7 Effect of irrigation on bulb quality

- Irrigation is one of the most crucial inputs for bulb crop production.
- Water serve as limiting factor for low productivity in bulb crop.
- The shortage of irrigation at bulb development usually coincides with summer season where irrigation replaces daily crop evapotranspiration.
- Irrigation level had exerted significant effect on neck thickness of bulb crop.
- The irrigation levels shows maximum neck thickness of bulb crop as IW/CPE=1.00 while minimum bulb crop neck thickness as IW/CPE=0.50.
- Maintaining consistent soil moisture with irrigation restricts excessive root growth and reduces risk of bulbs splitting.
- The net irrigation requirement is the amount of irrigation water required to be delivered in the field to meet the consumptive requirement of crop and other needs like:

- leaching,
- pre-sowing
- nursery water requirement

Thus, NIR (net irrigation requirement) = CIR + LR + PSR + NWR

Where

- LR = Leaching requirement
- PSR = Pre-sowing requirement
- NWR = Nursery water requirement

## 5.8 Effect of irrigation on bulb crop storage quality

- Water and nutrient management for bulb crop influence the storage behaviour of bulb directly or indirectly.
- Bulbs grown under low soil moisture regime are smaller which tend to loose more moisture and gets dried earlier during storage.
- small size bulbs with high surface area loose more moisture as water vapour losses occurred lengthwise from side of bulb crop.
- As nitrogen has adverse effect on storability of bulb due to rotting and sprouting, so irrigation and nutrients application needs to be planed to get higher yield and better shelf life.
- Micro sprinkler irrigation with better water application efficiency provides opportunity to create optimum soil moisture regime for shallow rooted crops for better maturity and storage life.
- The physical loss in weight of bulb crop was high when stored in gunny bags as compared to bulb stored on cemented flour as gunny bags were kept on stakes where air circulation was more as compared to bulb kept on cemented floor.

## 5.9 Stop irrigation before final harvesting

In bulb crops, irrigation:

- is required immediately after transplanting particularly in rainy season when temperature during transplanting is very high.
- in kharif season should be around 8-10 times, depending on rains.
- in late-kharif crop, should be around 12-15 times.
- in rabi seasons should be around 15-20 times.
- during bulb formation is necessary where moisture stress at during the stage results in low yield.
- should be frequent and light at weekly intervals for good bulb development and in increasing yield.
- such as sprinkler and drip, increases crop yield significantly.
- during rabi season should be stopped when tops mature and start falling.

- in kharif season should be stopped 10 days before harvesting.
- given to crops should be last given 10-15 days before harvesting in order to have low storage losses.

## 5.10 Irrigation scheduling

### 5.10.1 Introduction

- Irrigation scheduling is one of the factors that influence the agronomic and economic viability of small farms.
- It is important for both water savings and improved crop yields.
- The irrigation water is applied to the cultivation according to predetermined schedules based upon the monitoring of soil water status and crop water requirements
- The type of soil and climatic conditions have a significant effect on the main practical aspects of irrigation, which are the determination of how much water should be applied and when it should be applied to a given crop.
- In addition to the basic factors relevant to preparation of irrigation schedules examined, other important elements should also be considered like crop tolerance and sensitivity to water deficit at various growth stages and optimum water use.

### 5.10.2 Factors

There are many factors which bulb crop grower needs to consider when developing an irrigation schedule for bulb crops which includes:

- water requirements of plants in each zone
- local climate and weather
- soil composition
- sun exposure
- slope
- depth of root zone

#### Crop Water Requirement

- The crop type and size impact how much water is required to remain healthy in landscape.
- Information regarding crop water requirements for common landscape crops is provided by local utilities or extension offices or online.
- Immediately following crop installation, landscapes at their early stages require frequent irrigation intervals to establish the plants.
- Once landscape is established, the schedule should be adjusted to reduce the irrigation frequency appropriate for those crop/plants.

#### Weather Conditions

- Local climate and variations in evapotranspiration play a significant role in determining how irrigation schedules should shift over the season.

- Specific weather events like rainfall, should be considered and if scheduling is done manually, hen crop grower should take such parameters into consideration and create monthly schedules.
- A variety of irrigation controllers and sensors on market uses local conditions to determine watering times and amounts.
- Water Sense labeled weather-based irrigation controllers have been independently certified for efficiency and performance as they use local weather and landscape conditions to tailor watering schedules to actual conditions on the site.
- Soil moisture-based control technologies use moisture sensed in soil to adapt scheduling to plant water needs.
- Sensors like rain, wind and freeze sensors, helps in delay or in preventing irrigation, if conditions are not ideal or necessary for watering.

## Soil Types

- Soil type should be considered when developing irrigation schedule.
- The infiltration rate of water depends on composition of soil.
- Coarse soil requires frequent irrigation intervals; medium loamy soil requires longer, less frequent watering while fine clay soils have high water-holding capacity that requires the least frequent irrigation.
- If soil type is not taken into account when developing a schedule, water could be wasted due to runoff or deep percolation.
- Irrigation water on clay soil does not infiltrate quickly, and if system is run too long, landscape will begin to flood which results in runoff.
- To make irrigation in different soil types, include cycle and soak in irrigation schedule instead of irrigating each zone fully by using cycle and soak method irrigates in short intervals.
- If flow rate is slow enough, there will be no pooling on surface during entire cycle and hence there will be no need to adjust the schedule.

## Slope and Sunlight

- The slope of landscape and its orientation to sun should be considered while developing irrigation schedule.
- Bulb crop grower should take care that there should not be long irrigation schedule events on slopes, as it could result in water runoff the landscape.
- Ponding results at lower end of landscape due to runoff from elevated areas, so lower-flow-rate emitters should be installed, and/or more frequent, shorter events should be scheduled.
- Regarding sun exposure, crop/plants exposed to intense sunlight, or on slopes with southern or western exposure, may need more irrigation due to increased evapotranspiration of crop/plants.

### Depth

- The depth of root zone should be considered while scheduling the irrigation.
- If irrigation does not reach the full depth of root system, crops are not receiving enough water to maximize growth.
- Both plant type and age should be considered when determining root depth and associated schedule parameters.

### 5.10.3 Irrigation Scheduling for Onion

- Water use of onions varies throughout growing period and with weather conditions.
- The peak water demand for onions can be as high as 1.5- 2 inches per week.
- Peak use of water by onion crop occurs during latter stages of bulb enlargement during warm weather.
- Water gets transplanted by the onions very soon after setting, so use of 1/2-inch during the time, will help in making good contact between soil and roots which lead to good stand.
- During 2-3 months, plants is small which have relatively shallow root system.
- The fall months tend to be dry where irrigation is required when soil becomes dry in top 6 inches.
- Irrigation given during the period should be limited to about 1/2-inch per application, where irrigation applications are infrequent, as plants are small and water demand is low.
- When bulb enlarges, water demand will gradually increase and will require irrigation when weather becomes dry.
- Rooting depths during such stage will be around 12 inches or less, so irrigation applications should not be more than 1 inch.
- Typical applications should range between 0.6-1 inch for loamy soils and for sandy soils.
- During dry weather, irrigation should be done twice or thrice per week during warm weather.
- When temperatures become cool, irrigations should be less frequent.
- With many other bulb crops, onions do not generally wilt when they experience moisture stress.
- As moisture stress is difficult to detect by inspection, so crop grower should monitor the soil moisture by installing tensiometers or electric resistance blocks or other moisture sensor in soil.
- Crop grower can install the soil moisture sensors at two depths, one near the middle of root zone while other near the bottom, but best practice is to install one at 4-6 inches while other at 10-12 inches.
- The ideal range for soil moisture is between 5-20 centibars for most plain soils.

- ❖ If the system needs 3 days to water the entire field, then crop grower should install at least three soil moisture stations, evenly spaced around the field where each station should contain two sensors, one shallow and one deep.
- ❖ Crop grower should monitor the readings on the soil moisture sensors at least thrice in a week when the weather is dry.

### 5.10.4 Irrigation Scheduling Methods

- ❖ Irrigation requires relative high investment in terms of equipment, fuel, maintenance and labor, but it offers significant potential for increasing net income.
- ❖ Frequency and timing of water application have main impact on yields and operating costs.
- ❖ To schedule irrigation for most efficient use of water and to optimize production, crop grower should frequently determine the soil water conditions throughout root zone of crop being grown.
- ❖ There are many methods which have developed and are used that compares investment in irrigation equipment.
- ❖ When properly used and coupled with grower experience, a scheduling method improves irrigator's chances of success.

1. Water Balance Method

- ❖ In water balance method, the idea is to get balance of incoming and outgoing soil water so that adequate available water is maintained for the plant.
- ❖ In this, input includes incoming water in any form whether rainfall or irrigation while output includes any type of water removal which is known as evapotranspiration.
- ❖ Evapotranspiration consists of water removal by plant and water loss due to evaporation which uses two variations of water balance method.
- ❖ To use variation, crop grower must know the soil type and available water holding capacity of soil which is obtained for soil from the soil conservation guide.
- ❖ The grower should determine the zone which are trying to manage where zone varies as per effective rooting depth of particular crop.
- ❖ In this, effective rooting depth of certain crops maybe less, so determining the total water, grower have available to manage in such zone.
- ❖ It is desirable to try to manage only percentage of total water.
- ❖ As water is removed daily, such amounts are subtracted from adjusted water available column.
- ❖ When water available approaches zero balance, the grower should irrigate where the amount added depends on soil type which will be same as 50%.

2. Soil Moisture Measuring

- ❖ Many devices like tensiometer, electrical resistance meters are available for monitoring or measuring soil moisture.

- ❖ The most common in use in field are tensiometers, moisture blocks and various electronic soil probes.
- ❖ The expense of device results as a factor, so it is important for crop grower to make sure of all necessary inputs.

3. Computer Models

- ❖ Computer models are developed and updated each year.
- ❖ To use computer model, grower should have access to computer on regular basis.
- ❖ Simple models are available that do checkbook type record keeping.
- ❖ Many sophisticated models are present that needs weather data inputs.
- ❖ Models also incorporate other management decisions like whether to spray for insects and diseases.
- ❖ Models of later type are known as expert systems.
- ❖ Computer models are beneficial in decision making, but Garbage in – Garbage out is applicable in computer models.

4. Keeping a Chart

- ❖ Crop grower can get benefit of using tensiometers, electrical meters or computer models by recording soil moisture readings and plotting them on a chart.
- ❖ The grower can plot the reading directly in the field.
- ❖ Grower should use different colored pencils for different depth tensiometers for making chart easier to read where lines show activities in past.

# Chapter 6

# Integrated Pests and Disease Management in Bulb Crops

## 6.1 Identify stages of crop for pest and disease incidence

### 6.1.1 Introduction

Diseases and pests are the main factors that contribute to loss of harvest in crops. The resulting losses are to significant extent the consequences of crops grown in monocultures and continuing evolution of new races of pests and pathogens which are able to overcome resistance genes, that is introduced by modern breeding.

Pest identification is the initial step in pest management situation which depends on field scouting or monitoring of pest populations/occurrence and crop development. It is critical since appropriate management methods may vary dramatically for each pest. Unfortunately, pest identification can be very difficult if you are not familiar with the weed, insect, or disease problems present within an area. In addition, many pests share similar traits, which makes identification extremely difficult.

Field scouting can be accomplished in many ways including entering a field from several different points and surveying into the center, or subdividing it into several small sections and making observations within each section. However, representative sampling of the whole field is important because pests tend to congregate in specific locations.

Identification is significant to distinguish between insect pests and beneficial. Insect identification depends on morphological features like structure of mouthparts, wings, legs, antennae, etc. Some special equipment is required for effective scouting such as sweep net, forceps, and aspirator for collecting samples, vials containing rubbing alcohol used for killing and preserving collected specimens, and magnifying lens for identification of specimens. The type of damage observed in the field and where the pest is located on the plant will also help determine which pests are present.

Many pests have chewing mouthparts and eat plant tissue. Caterpillars of many butterflies and moths as well as larval or adult stages of several beetles feed on leaves, fruit, roots or other specific plant parts. Most plants are also hosts to one or more species of aphids, leafhoppers or plant bugs. These insects have sucking mouthparts for puncturing plant tissues and sucking out sap. This causes damage to leaves, flowers or fruit. Sucking insects can also cause crop losses by spreading diseases from infected to healthy plants. Typically, different pest species will attack a crop at a specific time of year or stage of plant growth or under specific environmental conditions.

Several types of microorganisms can cause a reduction in plant health including fungi, bacteria, virus, and nematodes. Identification of these organisms in the field is usually very difficult, and lab identification is often required. Tools needed to aid in locating and determining pathogens are a sharp knife, a shovel, and a magnifying lens.

## 6.2 Symptoms and damage of disease

### 6.2.1 Onion Diseases

**Fungal Diseases:**

Black Mold

- Black discoloration on the neck before harvesting; lesions on outer scales; black streaks on the outer dry scales; the entire bulb surface will turn black and shriveling.
- The Fungus is the main cause of this disease. Wash hands thoroughly after contacting affected parts of the plants.

Botrytis leaf blight:

- It results in high humidity and warm temperatures.
- It is a fungus which survives on piles of crop debris or in soil, older leaves and are highly susceptible to blight than younger leaves
- It shows presence of small white lesions with light green halos on leaf which expand gradually and resulting into leaf blighting

Downy mildew:

- It shows pale spots or elongated patches on leaves with gray-purple fuzzy growth on leaf surface.
- It turns leaves into pale and then yellow.
- It makes the tips of the leaves to start dying.
- Cool temperatures and wet foliage are the root cause for this disease.

Fusarium basal plate rot:

- It causes curving, yellow or necrotic leaves.
- In this, tips of the leaves will wilt and moves downward.
- The bulbs is infected which will turn brown and watery with rot spreading from stem plates.
- It lead to discoloration of stems.

- ❖ The disease occur in moderate to high temperatures

Fusarium damping-off:

- ❖ Seeds will rot and are covered with mold.
- ❖ The tips of the roots will get discolored root and turn to pink, yellow, red or black.
- ❖ The seedlings will get swollen and will wilt and die
- ❖ Moist and wet soils will survive this fungal disease.

Pink Root:

- ❖ The roots turn to light pink roots and will turn darken and turn purple when disease progress.
- ❖ The roots turn transparent and water soaked which lead to weak plant.
- ❖ The seedling affected by pink root will die.
- ❖ Matured plants affected by pink root become stunted and produce undersized, shriveled bulbs.

Purple blotch:

- ❖ It shows small water-soaked lesions on leaves or stalk with white centers.
- ❖ It lead to large lesions with girdle leaf.
- ❖ It spoils the tissue between the lesions and the leaf tip
- ❖ Wet foliage and high humidity will favor the diseases.

Rust:

- ❖ It shows small white flecks on leaves and stems which turn circular or elongated orange pustules.
- ❖ In this, severe infestations spoil the leaves and stems completely.
- ❖ High humidity levels and low rainfall levels will favor the disease.

Smut:

- ❖ The stems will turn dark, thickened and bend downwards.
- ❖ Blisters are formed on the base of older plants.
- ❖ In this, lesion mature and gets covered in black powdery fungal masses.
- ❖ The growth rate of plants will get reduced.
- ❖ Plants lead to death within 3-4 weeks.
- ❖ Smut will remain in the soil for many years and is induced in plants through insects.

White rot:

- ❖ Yellowing of older leaves.
- ❖ Slow growth
- ❖ Falling of leaves
- ❖ Fluffy white growth at base of bulb.

**Bacterial Diseases:**

Leaf streak and bulb rot:

- ❖ Water-soaked, dark green oval lesions and streaks on leaves
- ❖ Burns at tips of leaves
- ❖ Dark spots on wrapper scales of bulbs
- ❖ Discoloration on inner scales.
- ❖ It mainly effects during winter and spreads due to rainfall.

Dwarf Onion yellow dwarf virus:

- ❖ It shows yellow streaks at base of first leaves.
- ❖ The leaves get flattened or crinkled
- ❖ Plant produces undersized bulbs
- ❖ Flower stalks turn yellow and twisted and produces small flower clusters.
- ❖ This viral disease is transmitted through aphids.

Pythium seed rot:

- ❖ Seed turn water-soaked and mushy and start to decompose
- ❖ Roots turn gray and water-soaked
- ❖ Seedling collapse and die
- ❖ Older plants gets infected with virus which will be severely stunted.
- ❖ High soil moisture and cool temperatures will encourage the development of diseases.

## 6.3 Major insect-pest and their management through insecticides

### 6.3.1 Insect Pests of Onion

Thrips

- ❖ These are small insects that ranges in color from yellow to black and are 1/16" in length.
- ❖ They spend the winter as adults in crop remnants, alfalfa, wheat, greenhouses and weeds along crop fields.
- ❖ Adult thrips lay eggs singly where nymphs feed on leaves and pupation takes place in soil.
- ❖ They have rasping mouth parts tthat can be used to tear open plant cells to feed on plant juices.
- ❖ Their populations grows in hot and dry weather.
- ❖ Heavy rain or overhead irrigation lowers the populations.
- ❖ In onions, feeding takes place in protected, succulent areas where new leaves are emerges, deep between leaf blades.
- ❖ It causes damage in form of silver lines, white patches, tip dieback, curling and twisting of leaves, slowed growth, reduced bulb size and yields.

- Thrips damage increases the occurrence of purple blotch which penetrates the plant by wounds that is caused by feeding.

Leek Moth

- The adults are small, reddish-brown with white triangular mark on centre of folded wings.
- The hindwings of moth are fringed and are pale to dark grey.
- Its eggs are iridescent white, 0.4 mm in diameter and difficult to detect.
- Its larvae are yellowish-green with pale brown head capsule with many grey spots on each abdominal segment.
- At maturity, larvae reach 13-14 mm in length.
- The reddish brown pupa is encased in loosely netted cocoon.
- Many cocoons are found on host plants.
- In this, larvae causes extensive damage by tunneling mines and feeding on leaf tissue and sometimes on bulbs.
- On alliums with flat leaves, larvae feed on top and inside the leaf material and bore by folded leaves towards the center of plant, making series of pinholes on inner leaves.
- Larval mines in central leaves become longitudinal grooves in mature plant.
- Leek moth larvae enter hollow leaves like those of onions and chives, to feed internally, creating translucent windows on plant surface.
- With this, affected plants appear distorted and are susceptible to plant pathogens.
- With this, damage is more prevalent near field perimeters.

Onion Maggot

- The flies spend the winter as small brown pupae in soil.
- The adults emerge in spring which travels considerable distances for host plants.
- Female flies seek out their host crop to lay eggs at base of stem.
- Cool, moist soil conditions favor survival of eggs, and soil temperatures over 95 F kill them.
- When soil temperatures in upper ½ to 1 inch are above 100 degrees F then soil temperature provides control.
- There are three generations each year.
- Emergence of such pest is similar to cabbage root maggot which coincides with blooming of common roadside weed, yellow rocket.
- When eggs hatch, larvae feed on roots and causes destruction of root system.
- In onions, newly hatched larvae crawl behind the leaf sheath and enters the bulb, and feed on roots, stem and developing bulb.
- Feeding damage encourages entry of soft rot pathogens.

### 6.3.2 Insect Pests in Garlic

Mites:

- Bulb mites are pests found on garlic bulbs and other alliums.
- They are off white in color, shiny and globular and are found huddled together under plants roots.
- Bulb mites reduce harvest and stunt plant growth.
- They live from one growing season to next, so it is required to rotate plantings by skipping allium varieties in successive years.

Leaf miners:

- Leaf miners are pest found on garlic bulbs, first as eggs lain within the leaf tissue.
- The hatchlings are tiny, off-white larvae that tunnels in leaves, leaving visible damage.
- As resulting leaf damage is main cosmetic, presence of leaf miners damage other leafy crops in garden.

Wheat curl mites:

- Severe infestations of wheat curl mite causes twisted, stunted leaf growth.
- Its main impact is on bulb.
- It cause cloves to dry out.
- The mites act as vector for Yellow Streak virus.
- They are so tiny and are invisible to naked eye.

Nematodes:

- It is an insidious garlic bug which lives and reproduces in garlic plants.
- These microscopic worm like pests eat all parts of stems, leaves and bulbs.
- It can live without water and survive for many years in soil.
- Nematode nibbling results in deformation, discoloration and tissue collapse of bulb.
- Their population continues to expand exponentially for years with no visible signs of damage.

Thrips:

- They are garlic plant's common pest.
- They suck the sap from the plants, gradually slowing growth and bulb production.
- If infestation is severe, entire commercial garlic fields can wilt and die.

### 6.3.3 Use of insecticides in managing bulb crop

- Pest management is important for getting good quality of bulb.
- Arthropod pests and diseases reduces crop yield in terms of quantity and quality.
- Many insect pests of bulb crops (onion, garlic, leek, shallot, and chive) reduces yield potential.

- Management of bulb crops pests depends on insecticide usage during crop growing period, while insecticide resistance results in control failures which threaten long-term viability.
- Thrips are serious pests on wide range of fruit, vegetable, flower and agronomic crops.
- Thrips attack number of vegetable crops mainly onion thrips resulting in economic damage of crops by affecting young leaves, buds and flowers.
- As many onion fields require protection against onion thrips for particular time as per their varieties, and if such thrips are not controlled, it will lead to damage of reduced bulb yield.
- Chemical pesticides are applied to combat plant pathogens in agricultural fields.
- Insecticides such as dimethoate with 0.06% and Methyl Demeton with 0.05% are required in correct proportion to control the thrips.
- To suppress thrips population, Cypermethrin at 60 g ai/ha while Carbosulfan wih 0.05% and Lambda cyhalothrin should be effectively used.
- For controlling thrips in onion and garlic, insecticide should be used in sprayer and should be sprayed at base of the crop/leaves with higher pressure and spray volume.
- Insecticide applications should be scheduled as per scouting observations which should be done by randomly sampling of leaves or whole plants from different locations across the field.
- There are many insecticide products as shown that are used to control thrips which differs in characteristics related to impact on bees and beneficial insects, minimum re-entry interval, minimum pre-harvest interval, maximum number of applications per season, required frequencies of application and availability and usage restrictions.

| Active Ingredient | Products | Mode of Action | | REI (hours) | PHI (days) |
|---|---|---|---|---|---|
| | | # | Group | | |
| spinosad | Entrust®, Success® | 5 | spinosyns | 4 | 1 |
| spinetoram | Radient® SC | 5 | spinosyns | 4 | 1 |
| spirotetramat | Movento® | 23 | tetramic acids | 24 | 3 |
| abamectin | Agri-Mek® SC | 6 | avermectins | 12 | 30 |
| methomyl | Lannate® LV | 1A | carbamates | 48 | 7 |
| diazinon | Gowan Diazinon 4E | 1B | organophos-phates | 72 | 14 |
| acetamiprid | Assail® 30SG | 4A | neonicotinoids | 12 | 7 |
| cyantraniliprole | Exirel™ | 28 | diamindes | 12 | 1 |

REI – restricted entry interval, the number of hours from treatment until the area can be entered without protective clothing.
PHI – preharvest interval, the number of days required from treatment to harvest.
Based on 2016 recommendations from California, New York, and the Midwest.[1,3,4]

## 6.4 Repeat the spray as per demand

### 6.4.1 Sprayer

#### Introduction

- Crops of different types are widely affected due to pests and insects, many chemicals like pesticides and insecticides are prepared for controlling this.
- Is a machine used to apply liquid chemicals on plants to control pest and diseases.
- It can also be used to apply herbicides to control weeds and to spray micro-nutrients to enhance plant growth.
- For spraying chemicals, sprayers acts as suitable agriculture plant protection equipment.
- It helps in spraying pesticides with even distribution and regulates the amount of pesticide to avoid excessive application that might prove wasteful or harmful.

#### Functions

- The main functions of sprayer includes:
- Breaking chemical solution into fine droplets of effective size.
- Distributing droplets uniformly over the plants.
- Applying chemicals with sufficient pressure for positive reach of plants.
- Regulating the amount of liquid applied on plants to avoid excessive application.

#### Qualities

A sprayer posses following qualities:

- It produce steady stream of spray material in desired droplet size so that the plants to be treated may be covered uniformly.
- It deliver the liquid at sufficient pressure so that the spray solution reaches all foliage and spreads uniformly over the plant body.
- It is light weight but sufficiently strong, easily workable and repairable.

#### Parts

The sprayer comprises of:

- Nozzle body - It is the main component on which other component of a nozzle fit.
- Swirl plate - It is the part of a cone nozzle which imparts rotation to liquid passing through it.
- Spray gun - It is a lance from which spray is readily adjustable during the operation.
- Spray boom - It is a spray lance with spray nozzles fitted to head, that is kept at right angles to the lance.
- Filter - It is a component which is used to remove suspended matter larger than predetermined size from fluid.

- Over-flow pipe - It is a conduit through which excess fluid from a pump is by-passed by the action of a relief valve or pressure regulator.
- Relief valve - It is an automatic device to control the pressure of fluid or gas within range a predetermined value.
- Pressure regulator - It is an automatic device to control the pressure of fluid or gas within a range of settings.
- Cut-off valve - It is a manual mechanism between the pump and the nozzle which controls the flow of liquid from the sprayer.
- Nozzle disc - It is component containing the final orifice of a nozzle usually a cone nozzle.
- Nozzle boss - It is a lug on spray boom or spray lance to which a nozzle body or cap is screwed.
- Nozzle tip - It is component containing the final orifice of a nozzle usually a fan nozzle.
- Spray lance - A hand-held pipe through which the liquid reaches the nozzle mounted at the free end.

**Preparing the sprayer**

- Before spraying operation is started, rinse out the sprayer; remove and clean all nozzles, nozzle screens and strainers.
- Make sure strainers and nozzle screens are 50 mesh or larger when wettable powders are used.
- All nozzles should be of same type, size and fan angle for most applications.
- Check all lines, valves, seals and the tank after filling the sprayer with water and during calibration to be sure there are no leaks in the spray system.
- For safety, replace weather-cracked or worn hoses.
- Adjust the nozzle height and spacing as suggested by the nozzle manufacturer or as specified on the pesticide label.
- Only water that is clean enough to drink should be used in the sprayer.
- A small amount of silt or sand in the water rapidly wears pumps and other parts of the sprayer system.
- Operating a pump dry or with restricted inlet may damage the pump, so do not operate pumps at speeds or pressures higher than the manufacturer recommends.
- Power take-off pumps should be restrained from rotating by chains or torque bars.
- Do not use ground speeds that are too high for existing field conditions.
- Booms bouncing up and down or back and forth can cause application rates to vary by 50 percent, so such bouncing can damage the spray booms or frame of sprayer.

### 6.4.2 Using sprayer as required for bulb crop

- During cultivation or storage seasons, many diseases get transferred by flying aphid.
- When bulb is infected with virus, it transfers to offspring, so crop grower prevents the infection on crop as and when it is required.
- Insecticides are used for killing pests such as insects and arachnids which comes in aerosol and is sprayed on insect or its nest for killing.
- Spraying of insecticides depends on volume of insecticide formulation that is applied on given field is based on size of field, availability and cost of labour, price, availability of water tanks and/or pumps near field, height of crop canopy and systemic properties of insecticides.

## 6.5 Precaution during spraying of insecticide

### 6.5.1 Introduction

- Spraying is usually done by means of spray gun or trolley.
- Pesticides are chemicals that deter and eliminate certain pest populations including insects, rodents, fungi, weeds, and other animals.
- The increased quantity and frequency of pesticide applications have posed major challenge to targeted pests causing them to either disperse to new environment and/ or adapt to the novel conditions.
- The adaptation of pest to new environment could be attributed to the several mechanisms such as gene mutation, change in population growth rates, and increase in number of generations etc.
- Pesticides have also been reported to influence mineralization of soil organic matter, which is a key soil property that determines the soil quality and productivity.
- Pesticides that reach the soil may also disturb local metabolism or can alter the soil enzymatic activity.

There are certain situations where crop grower can be exposed to insecticides which happens during:

- Delivery or storage of chemicals
- Preparation, transportation or application of spray solution
- Re-entry of greenhouse and crop after application
- Cleaning and repairing of spray equipment
- Disposal of empty containers

### 6.5.2 Precautions for applying insecticides:

- Select proper spray equipment to avoid any leakage.
- The applicator should not blow/clean clogged nozzle with mouth, but should use tooth-brush to clean it.
- Apply only at recommended dose and dilution.

- Spraying should be done at time when wind velocity is low and should be in proper direction.
- Never apply spray solution without wearing protective clothing.
- Personal protective equipment prevents insecticides from coming in contact with body or clothing or being inhaled.
- The protective equipment includes clothing which covers arms, legs, nose and head while working with insecticides.
- Gloves and boots are used to protect the hands and feet while hats, helmets, goggles and face masks protect the hair, eyes, and nose.
- Respirators are used to avoid breathing dust, mist or vapour.
- Wear gloves while doing spraying as the solution may cause damage to hand.
- Emulsifiable concentrate formulations should not be used for spraying with battery operated ULV sprayer.
- Living organisms should be kept away from the area where spraying has been done.

## 6.6 Perform proper ploughing and pre irrigation for breaking down of residue (if any)

### 6.6.1 Land levelling

- Land leveling is an important agronomic practice which removes up and downs of soil and to get better seed bed conditions for uniform distribution of irrigation water and better supply of other inputs to grow crop.
- Crops need fertile loam to clay loam soils with good drainage for better yields.
- The crop growing areas should be divided into small plots of half an acre along with channels in the middle to separate each plot where proper use of irrigation water should be there for crop needs.

### 6.6.2 Land Preparation

Bulb crop needs well crushed but compact seed-bed for good and uniform seed germination. In irrigated areas, after harvesting of previous crop, field should be ploughed with disc or mould board plough.

The idea of land preparation is to provide necessary soil conditions to enhance successful establishment of young offshoots or tissue culture plants received from nursery.

The aim of preparing the land before sowing involves:

- Creation of soil structure which encourages the seedlings to emerge rapidly and uniformly and allowing young plants ready access to vital resources of nutrients, water and aeration.
- Full tillage and minimal tillage systems where minimal tillage limits soil disturbance the surface layers or just shallow slot for each crop row.
- Minimal tillage usage that can be taken into the field when the soil is wet, as perhaps after a previous rice crop, it has the prime advantage that it commonly shortens the time between successive crops/rotations which can be critical for maximise yield per year.
- Incorporating any additives like lime, compost, farmyard manure and chemicals for plant nutrition and pest control depending on location that incorporate residues remaining from previous crops.
- Controlling of weeds, pests and diseases.
- Shaping of land so that irrigation water can be applied efficiently and drained effectively and so that water logging is avoided or minimised.

### 6.6.3 Land preparation operations

Once the required area for establishing the plantation is selected by wheat grower and planning operation has been finalised, the actual preparation of land can be started which can be divided to structure and pace the implementation process in order to be ready for planting at the most suitable time, according to the specific regional climatic conditions.

#### Mechanical field preparation

The mechanical or initial soil preparation concerns mainly the preparation of a field for further detailed preparation such as irrigation system installation, hole preparation, if applicable to the area, include:

(i) debushing/bush clearing;

(ii) removal of stones and rocks;

(iii) ripping;

(iv) levelling of soil.

#### Pre-irrigation

- Pre-irrigating involves flooding the field for 2 to 3 days before the first plowing to moisten the soil and facilitate plowing.

#### Irrigation system installation

- The type of irrigation system to be used will be determined by the availability of water, topographical and soil conditions.
- When the initial soil preparation is completed, the installation of the required irrigation system will be implemented according to the prescribed design.

### Ploughing

Ploughing is main tillage operations which is done to cut, break and invert soil partially or completely suitable for sowing seeds. It is used for:

- ❖ special technologies for problem soils
- ❖ obtaining deep seed bed of good texture.
- ❖ increasing water holding capacity of soil.
- ❖ improving soil aeration.
- ❖ destroying weeds, insects and pests.
- ❖ adding fertility to soil by covering vegetation.

### Planking field after ploughing

- ❖ Planking is done to crush hard clods to smoothen soil surface and to compact soil lightly.
- ❖ The field is made ready for sowing after ploughing by harrowing and planking.
- ❖ Planking should be done early in morning.
- ❖ A buffer zone of 10-30 m width should be kept to avoid contamination from the non-organic field.

## 6.7 Use of different insect traps for control of pest

### 6.7.1 Insect trap

- ❖ It is used to monitor or reduce populations of insects/arthropods by trapping and killing them.
- ❖ It uses food, visual lures, chemical attractants and pheromones as bait and are installed so that they do not injure other animals or humans resulting into residues in foods/feeds.
- ❖ It varies in shape, size and construction and reflect the behavior or ecology of target species.
- ❖ It includes light traps, adhesive traps, flying insect traps, terrestrial arthropod traps and aquatic arthropod traps.

### 6.7.2 Different insect traps

#### Light traps

- ❖ These are set up at night which attracts different flying insects such as moths, mosquitoes, chafer beetles, american bollworms, army worms, cutworms, brown rice plant hopper, green rice leaf hopper, rice black bugs, rice gall midges, rice stem borers and tomato hornworms.
- ❖ It is normally constructed using bamboo/wooden poles and are arranged as tripod which are set downwards in ground to secure from wind/animals.
- ❖ In this, lantern is suspended from top over the bowl of water with less oil in it.

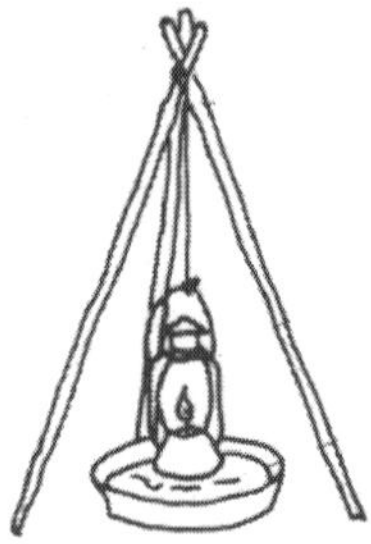

- ❖ The best time for placing light traps around garden/field depends on life cycle of insect and development stage of crop.
- ❖ The correct time of installation is soon after the moths emerge and before laying of eggs.

Fly trap

- ❖ It is a large board that measures 30cm x 30cm which are painted with bright yellow/orange and is covered with adhesives such as oil or glue.
- ❖ There are different pests that are attracted to different colours as flies get attracted to bright colour of board and flies on it which gets stuck in oil or glue and dies.

Pheromone trap

- ❖ Pheromone is sexual attractant which is produced by female insects.
- ❖ If trap is bait with this, it attracts male insects in trap from which they cannot escape.
- ❖ This trap is prepared by commercial companies and is costly to farmer/grower.

Adhesive trap

- ❖ It is a sticky trap which has simple flat panels or enclosed structures, that are baited, that ensnare insects with adhesive substance.
- ❖ Baitless has nickname as blunder trap, where pests blunder in them while wandering or exploring.
- ❖ It is widely used in agricultural and indoor pest monitoring.
- ❖ Baited shelter trap like Roach Motel and similar enclosures have adhesive material to trap the insects.

Terrestrial arthropod traps

- ❖ Pitfall traps are used for ground-foraging and flightless arthropods like Carabid beetles and spiders.

- Pitfall traps has bucket or container in soil or other substrate so that its lip is flush with substrate.
- Grain probe trap is used to monitor pests of stored grain having long cylindrical tube with many holes along its length that can be inserted at many depths in the grains.
- Soil emergence traps has inverted cone or funnel with collecting jar on top which are used to capture insects with subterranean pupal stage.
- Emergence traps monitor important disease-vectors like Phlebotomine sandflies.

## 6.8 Identify early symptoms of various types of diseases

### 6.8.1 Diseases in Leek

Botrytis leaf blight Botrytis squamosa

- It is a fungal disease which shows small white lesions with light green halos that expands during aging.
- In prolong periods of moisture, fungus develops rapidly and cause leaf blighting.
- The disease emergence favors high humidity and warm temperatures.
- With this, fungus survives on piles of crop debris or in soil.
- With this, older leaves lead to more susceptible to blighting as compared to younger leaves.

Damping-off Fusarium spp.

- It shows discolored root tips which can be pink, tan, yellow, red or black.
- It leads to slow growing of seedlings which will wilt and die
- In this, fungus survives in soil and diseases emergence due to moist to wet soil

Downy mildew Peronospora parasitica

- It shows pale spots or elongated patches on leaves.
- It shows gray-purple fuzzy growth on leaf surface which turn leaves to pale then yellow and finally collapsing of leaf tips.
- The disease is spread in cool and humid weather

Pink root Phoma terrestris

- It shows light pink roots which darken and turn purple.
- In this, roots become transparent and soaks water.
- It lead to nutrient deficiency in plant.
- With this, the infected seedling dies leading to stunted plants with undersized and shriveled bulbs.
- It is a fungus which colonizes plant by root tips that survives in soil uptu depth of 45 cm.

Purple blotch Alternaria porri

- It shows small water-soaked lesions on leaves or stalk with white centers that enlarges to become zonate and turns brown to purple with red or purple margin around yellow zone.
- The large lesions coalesce and girdle leaf which kills tissue between lesions and leaf tip.
- The severely infected foliage results in dying.
- The disease emerges due to wet foliage using sporulation that takes place during night and in high humidity.

White rot Sclerotinia cepivorum

- It leads to old leaves to turn yellow.
- It shows stunted growth and death of all leaves.
- It shows fluffy white growth on base of bulb that spreads up bulb to store leaves.
- It is a fungus which survives in soil for 20 years and result as most damaging diseases of crops worldwide.
- It leads to major crop losses.

### 6.8.2 Diseases in Onion

Black mold Aspergillus niger

- It is a fungal disease that shows post-harvest black discoloration at neck and lesions on outer scales.
- It shows black streaks under outer dry scales where entire surface of bulb turns to black and shrivel.

Botrytis leaf blight Botrytis squamosa

- Small white lesions with light green halos which may expand slightly as they age; in prolonged periods of moisture fungus may develop rapidly and cause leaf blighting
- Disease emergence favors high humidity and warm temperatures; fungus survives on piles of crop debris or in soil; older leaves more susceptible to blighting than younger leaves

Downy mildew Peronospora destructor

- Pale spots or elongated patches on leaves; gray-purple fuzzy growth on leaf surface; leaves turning pale then yellow; leaf tips collapsing
- Disease emergence favored by cool temperatures and leaf wetness

Fusarium basal plate rot Fusarium oxysporum

- Curving, yellow or necrotic leaves; necrosis begins at leaf tips and moves downward; wilting plants; infected bulbs may be brown and watery with rot spreading from stem plate to basal leaves; stem plates may have brown discoloration
- Disease emergence favors moderate to high temperatures

Fusarium damping-off Fusarium oxysporum

- ❖ Rotting seeds that are covered in mold; discolored root tips which may be pink, tan, yellow, red or black; slowly growing seedlings which wilt and die
- ❖ Fungus survives in soil and disease emergence is favored by moist to wet soil

Pink root Phoma terrestris

- ❖ Light pink roots which darken and turn purple; roots become transparent and water soaked; plant may look like it has a nutrient deficiency; infected seedling may die; stunted plants with undersized, shriveled bulbs
- ❖ Fungus colonizes plant through root tips; fungus can survive in soil down to a depth of 45 cm.

Purple blotch Alternaria porri

- ❖ Small water-soaked lesions lesions on leaves or stalk with white centers; which enlarge to become zonate and brown to purple in color with red or purple margin surrounded by yellow zone; large lesions may coalesce and girdle leaf, killing any tissue between the lesions and the leaf tip; severely infected foliage may die
- ❖ The disease emergence favors wet foliage with sporulation during night and in high humidity

Rust Puccinia porri

- ❖ Small white flecks on leaves and stems which develop into circular or elongated orange pustules; severe infestations can cause leaves to yellow and die
- ❖ Favors high humidity but low rainfall; spores can be transported over long distances by wind

Smut Urocystis colchici

- ❖ Dark, thickened lareas on cotyledons which become large and causes leaves to bend downwards; raised blisters may be present on the base of scales in older plants; lesion mature and become covered in black powdery fungal masses; plant growth stunted; death of plant occurs within 3-4 weeks
- ❖ Smut can persist in soil for many years and is mainly introduced through infected sets and transplants; spores can be spread by wind, on equipment or in irrigation water

White rot Sclerotium cepivorum

- ❖ It makes older leaves to turn yellow with stunted growth leading to death of all leaves.
- ❖ It shows fluffy white growth on base of bulb which spreads up bulb to store leaves.
- ❖ It is a fungus which survive in soil for 20 years and is one of most damaging diseases across world causing major crop losses.

Leaf streak and bulb rot Pseudomonas viridiflava

- ❖ It is a bacterial disease which soaks water.
- ❖ It shows dark green oval lesions or streaks on leaves causing tip burn of leaves.
- ❖ It shows dark spots on wrapper scales of bulbs and shows reddish-brown discoloration of inner scales.

- It lead to rot developing in ring-like pattern.
- The greatest damage occurs during winter and spreading of disease on infected plants more in rainy season

Onion yellow dwarf Onion yellow dwarf virus (OYDV)

- It is a virus which shows yellow streaks on bases of first leaves.
- It makes all leaves which emerge after infection to yellow leading to flattened or crinkled.
- The bulbs are undersized making flower to stalks yellow and twisted.
- The disease gets transmitted by many species of aphid, including peach aphid.

## 6.10 Use of resistant varieties of bulb crops

- Resistant varieties suppress or retard a pathogen's activity and show little or no symptoms of infection.
- Varietal resistance depends on field and laboratory observation as environmental conditions and local practices affects variety characteristics and performance, actual crop yields, quality and level of claimed pest and pathogen resistances on certain factors beyond control of varieties originator.
- Tolerant varieties do not significantly inhibit pathogen and shows severe disease symptoms in crops, without significant losses in yield or quality.
- As partial resistance/tolerance depends on many genes rather than single major one, it slows the spread of disease and impose less selection pressure on fungus with durability.
- Disease resistant varieties are attractive as they poses less or no risk to environment and enable growers to reduce and eliminate the need for pesticides.
- In some host-pathogen, resistance persist for many years, but in others it may be short lived.
- Disease-resistant varieties allow vegetable plants to grow and fruit when under pressure of a pathogen, its vectors and/or environmental factors that can promote infection and disease.
- Some plants can tolerate the presence of large numbers of insects without being severely affected. This is not very satisfactory however as insects will still cause damage, and in fact further breeding and population expansion of the pest species is supported.
- Due to light leaf spot and Phoma in autumn and spring seasons in crop, varietal resistance in crop should be taken into account.
- Bulb črop growers should test new varieties routinely to see which variety performs best on their fields, selecting those with tolerance or resistance to diseases like pink root and Fusarium basal rot if their fields have history of such diseases.

## 6.11 Undertake pruning of plant if affected by diseases (if need arises)

### 6.11.1 Pruning

- Pruning is a horticulture practice that involves selective removal of certain parts of plant like branches, buds or roots.
- Pruning helps a plant exhibit its most desirable attributes.
- Pruning of plants is done to remove deadwood, giving proper shape to plants, improving or sustaining health, reducing risk from falling branches, preparing nursery specimens for transplanting and harvesting and increasing yield or quality of flowers/fruits.
- Proper pruning improves the health and beauty of plants, thereby increasing their lifespan and productivity.
- Pruning reduces the amount of wood on plant, making more energy available for development of large, healthy flowers.

### 6.11.2 Pruning on onion

- Pruning is direct way of orienting different parts of plant for supply and distribution of food materials in vegetative or reproductive channels.
- Pruning showed influence on all the vegetative and quality parameters of onion cv.
- Pruning is done mainly for balancing and influencing nutrients and hormones.
- Due to pruning, existing plant face less competition for nutrients and hormones.
- Pruning associated with proper age of seedling is an important factor for successful onion production.
- As pathogen of leek, onion and garlic are hosts of virus as leeks are vegetatively propagated and leek yellow stripe spreads through pruning and handling of infected plants and bulbs.
- In case of disease in onion, crop grower should perform three levels of pruning which includes; Root pruning where roots are reduced, leaf pruning where leaves are reduced and leaf-root pruning for reducing leaves and roots.
- Leaf Pruning increases size of onion which results in high fruit yield, but better and stable fruit yield of onion results from pruning and gibberellic acid.

# Chapter 7

# Harvest and Post Harvest Management in Bulb Crops

## 7.1 Stoppage of irrigation 15 days prior to digging

- As onion has shallow root with restriction up to 3 cm on top and 15 cm depth has less water needs at initial stage as per crop growth, soil type and planting season.
- Onion requires one irrigation immediately after transplanting in rainy season where temperature during transplanting is very high.
- Atleast 7-10 irrigation is required for proper growth and good yield.
- If tillage is not done before sowing, use light irrigation in 2-3 days of sowing.
- Apply irrigation if moisture in soil reduces to 15-30%.
- Apply first irrigation after 8-10 days of sowing in sandy loam and loam soils and 10-12 days of sowing in heavy soil.
- If there is possibility of very low temperature or frost, irrigate immediately.
- Irrigation of bulb crop should be stopped when tops mature and start falling.
- In kharif season, irrigation should be stopped 10 days before harvesting as it gives lower storage losses.

## 7.2 Harvesting stage of onion crop in field

### 7.2.1 Harvesting

Harvesting is the operation of cutting, picking, plucking digging or combination of operations for removing crop from under/above ground and removing of useful part of fruits from plants.

- It can be carried out using:
- Slicing action with sharp tool.
- Tearing action with rough serrated edge.
- High velocity single element impacting with sharp/dull edge.
- Two elements scissors type action.

### 7.2.2 Harvesting stage

- Onion is harvested depending on the purpose for which the crop is planted.
- The harvesting of onion depends on proper planting and care all through growing season.
- For green onions, harvest the top as soon as they becomes 6 inches in height as longer top became strong which are difficult to harvest.
- In case of bulbs that are bolted or formed flower stalks, crop grower should pulled them as they are not good for storage.
- The time of harvest of bulb onion start with onion tops which falls over and turn brown in 100-120 days after planting as per cultivar.
- If weather is dry and there's no danger of frost, plants can be gently pulled out from the soil and can be laid right in the fields for day or two.
- If the weather is wet or frost is possible, move the onions immediately into protected area.
- Onion grower should see that good harvesting time of onion should be early in morning when temperature is not too high.
- Leeks should be pulled out when it is about an inch in diameter in 120-130 days.
- In mild climates, leek can be left in ground over winter and can be dug as required while in cold, they should be harvested before frost.

## 7.3 Diagnosis harvest based on leaf colour

- Onion plant has leaves which arise alternately from small flattened stem or base plate, so that older leaves are on outside while younger leaves are on inside of stem.
- In onion, each leaf is composed of a photosynthetic leaf blade and non-photosynthetic storage leaf base.
- During the growth of the plant the leaf scales thicken and form the characteristic bulb.
- During onset of bulbing, onion leaf sheaths swell and bladeless bulb scales gets initiated which swell to form main storage tissue of bulb.
- Leaf blades initiate before bulbing develops to full expansion and green top lodges at top down.
- When bulbs develop from leaf bases of onions are completely formed, then leafy green tops turns to yellow which eventually collapse at a point little above the top of bulb, making upright short neck.
- In kharif season, as onion tops do not fall, bulbs are harvested soon after the colour of leaves changes to slightly yellow and red pigmentation on bulbs develop.
- In garlic, leaves are flat, linear, gray-green and folded having keel on lower surface where 6-12 of them grows widely spaced along main stalk of the plant.

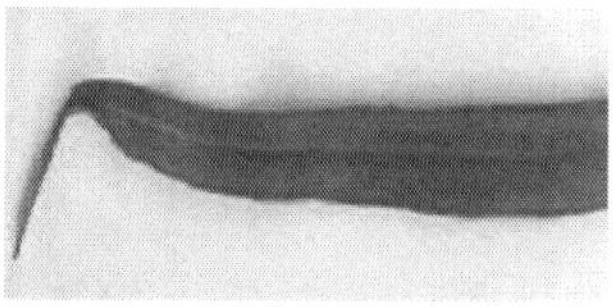

- In garlic, bases of non-topsetting types forms semi-stiff pseudostem, that remains upright till bulb matures, while bending near ground level.

## 7.4 Diagnosis of harvesting based on neck falling

- Thick-neck bulbs that result due to premature harvesting cannot be stored well.
- Onions should be harvested at 50% neck fall stage which gives indication of maturity.
- Onion bulbs reach maturity when plants cease to produce new leaves and roots.
- Harvesting decides storage life of onion as bulbs may be stored for about six months.
- If the neck of garlic bulb is hard, bulb grower needs to cut the garlic and should see for some air spaces between stem and cloves which shows maturity and if cloves has single mass, then there results early sign of maturity.

## 7.5 Implements used in harvesting

### 7.5.1 Introduction

- Onions, garlic etc. grows below the soil surface, as such crops are harvested by lifting the bulbs, tubers or pods from the soil before.
- Once lifting has been done, the crop is picked up by hand, sorted and packed.
- Lifting is done using specialized implements containing a blade, which is attached to back of a tractor.
- Manual or mechanical handling involves pulling of plants from soil and removing of tubers. Onion, garlic tubers can be placed in bulk trailers or bins on trailers and transported to pack houses whereas in onions, bulbs are left on field for at least week to cure before dry leaves are cut from bulbs and are kept in onion bags.
- For doing day-to-day routine work of crops, crop grower uses different tools, implements and accessories.
- Among these tools, some tools are simple and are used for simple operations while for doing specific operations, crop grower uses specific types of equipments.

### 7.5.2 Tools for land work

Kudali: It is an important tool used for digging of soil, pits or basic digging work before preparation of nursery beds. It has a metal blade attached to wooden handle and is operated manually.

**Fig Kudali**

Garden fork: It is used for breaking of soil clods and separation of medium size soil during preparation of nursery bed or fields.

Weeding fork: It consists of long handle with a blade of handle teeth used for loosens the soil and helps in weeding. It is drawn manually using the handle to collect the weeds and cops of plants etc.

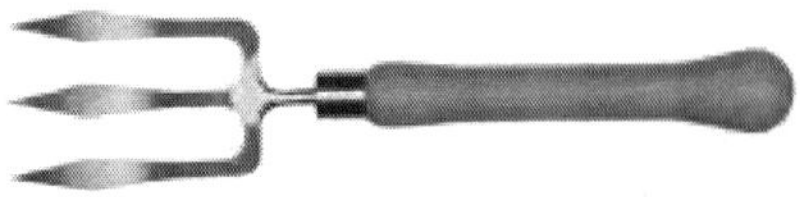

**Fig Weeding fork**

Crow bar: It is an iron rod with one end pointed and other as a wedge shaped. It is used for digging out large or hand boulders from the soil and digging of pits. It is also used for breaking hard soil pan.

Shovel: It has iron blade of spoon shape and wooden handle and is used in field transportation of dug soil for leveling of fields for preparation of nursery beds.

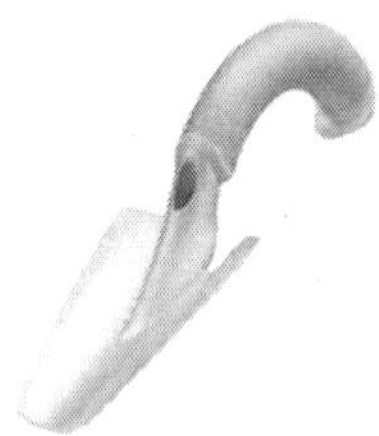

**Fig Shovel**

Pick-Axe: It is made of carbon steel having two edges with provision of axial hole for attachment with handle where one edge of pick axe is pointed while other edge is broadened. It is used for digging hard, compact and stony soils.

Spade: It is used for lifting and turning the soil and also for digging the pit, preparing channel for irrigation and drainage lines.

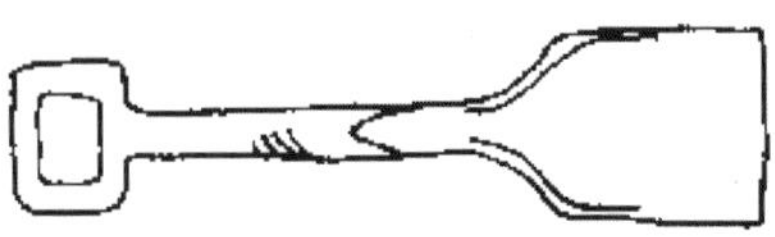

**Fig Spade**

Hoe-cum-Rake: It is rectangular shaped metal blade with fork like fore edge used for digging, hoeing, earthing, leveling and collecting weeds.

Furrow opener: It is used for opening narrow and shallow furrow after sowing seeds in nursery.

Hand leveler: It consists of narrow rectangular metal blade attached to long wooden handle.

Trowel (Khurpi): It is of shovel shape but small in size with iron blade and wooden handle. It can be made in many shapes as per local designs and requirements. It is used for hoeing, weeding and nursery plants and also for transplanting seedlings.

**Fig Trowel**

Axe: It has an iron blade which is fastened to wooden handle and is mainly used for felling trees and cutting branches.

**Fig Axe**

Bill Hook (Darat): It is made of iron curved at the far end which is used for cutting hardy branches of plant and other woody shrubs in the field.

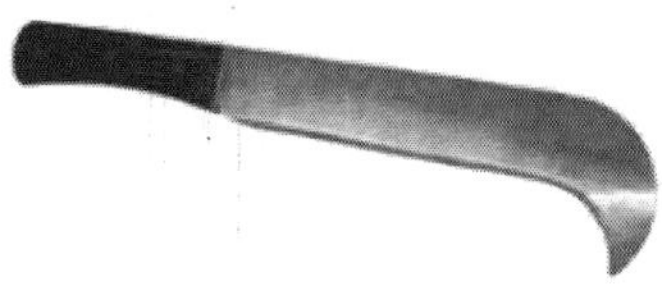

**Fig darat**

Sickle: It is used for cutting grass and leafy vegetables.

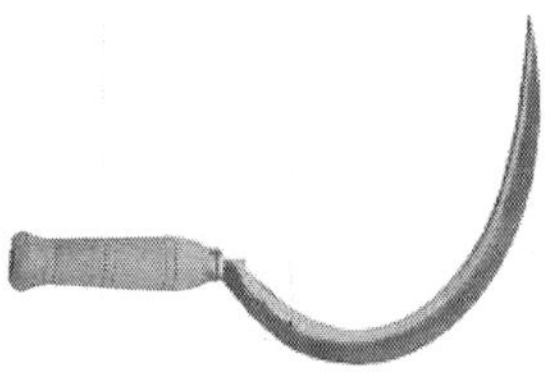

**Fig Sickle**

Wheel-Barrow: It is manually operated small trolly used for carrying nursery plants, compost, fertilizes, leaf litter, horticultural produce, stones etc. from one place to another place.

**Fig Wheel-Barrow**

Cultivator: It can be tractor or hand driven implement used for tilling the soil efficiently. It has tynes of quality carbon comprising of different sizes depending upon the plough depth.

**Fig hand driven implement**

Disc harrow: It is used for pulverizing the soil and for turning crop residues, weeds and other debris into the soil. Also it can be used for deep ploughing and turning of soil.

**Fig Disc harrow**

### 7.5.3 Mechanical harvesting machine

Machineries are required at every stage of crop production, starting from seed bed preparation, pit digging, nursery raising and transplanting, seed sowing, weeding, intercultural operations, spraying and harvesting as also to meet the specific requirements of small and marginal growers.

**Root Crop Harvester**

- It is used for digging different crops like onion and garlic on single bed of 1.1 m wide.

- The machine has a digger blade with its width as 1144 mm and thickness 16 mm.
- The blade of harvester was mounted on machine at an angle of 20° with horizontal.
- It has an elevator conveyor which is attached behind the blade.
- In this, the spacing between MS rods of elevator conveyor is around 20 mm with slope of elevator conveyor as 18°.
- The conveying system has two oval agitators which are used to separate soil particles from tubers.
- In this, power to elevator conveyor is given by gear box.
- The blade has two coulter discs at outer ends for easy slicing and lifting of soil.
- The extension at rear is given for increasing separation time of soil.
- The triangular blade of 1730 x 60 mm dimension is used for digging of onion using two wheels for depth control.
- The harvester has a field capacity of 0.20, 0.26, 0.22 and 0.24 ha/h for digging onion, garlic and other bulb crops.

**Tractor mounted vertical conveyor reaper**

- It comprises of reciprocating cutter bar assembly, seven crop row dividers, two vertical conveyor belts fitted with lugs, pressures springs, pulleys and gearbox for power transmission system.
- In this, crop row dividers are fitted in front of cutter bar assembly.
- The star wheels are mounted over the crop row dividers.
- The machine is kept in front of tractor where the power is given from tractor PTO using intermediate shaft that runs below the chassis of tractor and coupling shaft.
- Here, the height of machine which is above he ground is controlled by tractor hydraulic using pulleys and steel ropes.

- Once the crop is cut by cutter bar, it is held in vertical position and delivered to one side of machine by lugged belt conveyors which falls on ground as windrow perpendicular to machine movement direction.

### Tractor Operated Garlic Harvester

- In garlic harvesting, more labour is consumed for harvesting.
- Crop growers insist for single pass of duck foot cultivator for loosing of garlic bulbs from soil which are collected manually.
- In tractor operated garlic harvester/windrower, there is a triangular point knife blade which is mounted on MS flat for use in heavy soil.
- The harvester has chain type separating mechanism having MS bars of 12 mm at spacing of 38 mm along horizontal vibrating forks of 250 mm length at spacing of 200 mm for good separation of soil from clods.
- In this, windrower unit has MS bars at spacing of 25 mm which is kept at 900 to conveyor direction in unequal halves.
- In this, power is delivered from tractor pto to gearbox in 1:1 ratio.
- The gearbox with crank pinion arrangement gives power to oscillating vibrating fork.
- Tractor operated garlic harvester has an effective field capacity of around 0.26 ha/h.

## 7.6 Curing of bulb

- Curing is essential to obtain maximize storage life and have minimal decay.
- Onion bulbs are formed at soil surface in sandy soils where mature bulb gets pulled out by hand.
- Curing remove extra moisture from outer layers of bulb before storage as dry skin serves as surface barrier to water loss and microbial infection, and saves main edible tissue in fresh state.
- If onions or garlic are made into strings for storage or sale, it is good to separate them into sizes so that bulbs will be uniform in size in any string.
- In this, post-harvest treatment required for long storage of bulb onions is possible thorough curing of bulbs.
- Curing is a drying process which dries the necks and outer scale leaves of bulbs to prevent loss of moisture and attack by decay during storage which is done in fields under dry conditions through windrowing the bulbs.

- Under dry, warm conditions harvested onions are left in field for few days till green tops, outer skins and roots are fully dried.
- Under wet conditions, it may be necessary to dry onions on racks or trays under cover.
- For curing, heat and good ventilation with low humidity dries the neck and outer layers of the bulb.
- Outer layer that is contaminated with soil falls away when bulbs are cured which expose dry under-layer and if drying is not done in field, they can be collected in trays that can be stacked in warm, covered area with good ventilation.
- The curing of root crops replaces and strengthens damaged areas of skin, restoring protection against water loss and infection by decay organisms.
- Curing of dry bulb onions is done immediately after harvesting.
- In field curing, the bulbs moisture content decreases gradually from initial moisture.
- At maturity stages of 20%, 30%, 40%, and 50% of tops down, curing of onion bulbs reduces required time of curing by 68, 72, 76 and 82% for large size bulbs while 72%, 76%, 79% and 82% for medium/small size bulbs.

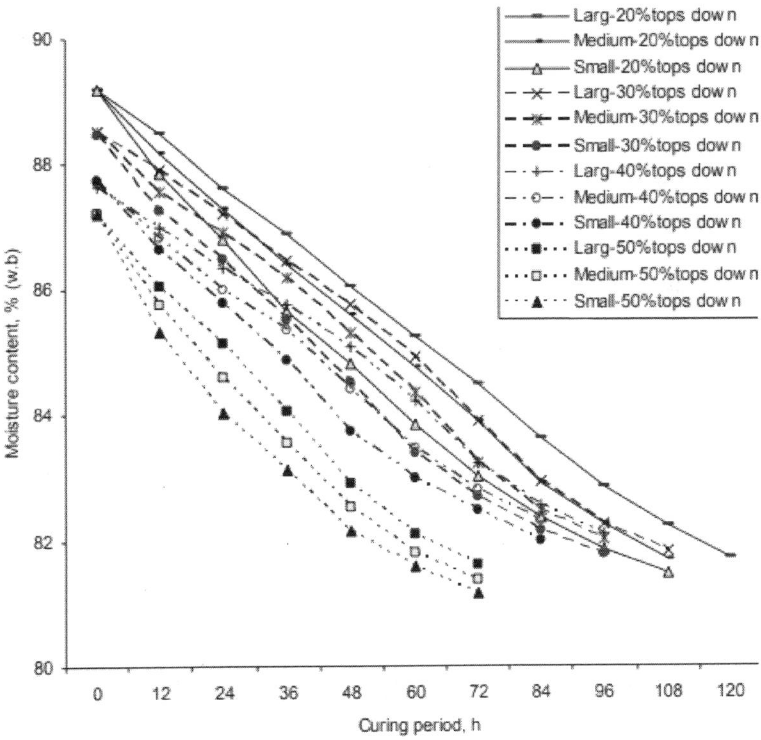


- Curing period of bulbs decreases as bulb size decreases from large to small and decrease as harvest stage changes from 20% to 50% tope down.
- Curing bulbs artificially harvested at 20% top down caused gradual decrease in bulb moisture content from initial to final moisture in 120 h for large bulbs and 108 h for medium and small size bulbs.

- At 30% tops down, decrease in bulb moisture reaches to 108 h for large bulbs, and 96 h for medium and small size bulbs.
- At all bulb sizes and harvest stages with 50% tope down, an extreme decrease in curing period of 72 h exists.

## Methods

Curing of onion bulb can be done by:

- windrowing,
- braiding,
- single layer or tray curing
- accelerated curing

Windrowing

- In small-scale operations, drying of onion is done in fields which involves harvesting the mature bulbs and laying them on their sides on surface of soil to dry for 1 or 2 weeks.
- In wet weather, bulb takes longer time to dry and develop higher levels of rots during storage.
- In this, sides of bulb will be in contact with wet soil or moisture which develops brown strains or pixels, that reduces appearance quality and value.
- Windrowing is weather dependent which cannot be relied on for large scale commercial onion production business.
- Bulbs harvested for its storage need 14-20 days of ripening or drying before being stored.
- In this, harvested bulbs are taken straight from the field and are dried artificially either in a store, shed, barns, or in purpose-built drier.
- It is commonly used when crops are stored in bulk that can also be applied to bags, boxes or bins.
- In this method, bulbs are laid on racks where heated air is passed across the surface of bulbs night and day.
- Drying here takes 7-10 days and can be completed when neck of bulbs have dried out and are tight and skins shriek when held in hand.
- The control of humidity level in store is critical and under high humidity, drying is delayed which lead to fungal increase in infection.
- If relative humidity is below 60%, excessive water loss and splitting of bulb makes outer skins to occur, resulting in storage losses and reduction of bulb value.
- Placing onions on wire mesh in well ventilated conditions and using air at about 30°C, 60-75% rh and 150 $m^3$ h-1.$m^{-3}$ is generally recommended for mechanical drying of onions.

## 7.7 Sorting and grading of bulb

Sorting and grading is done after storage to fetch better price.

### 7.7.1 Qualities of Bulb

The bulbs should be:

- uniform in shape, size colour and pungency of variety /type
- mature, solid in feel, reasonably firm with tough clinging skins.
- throughout cured and dried.
- free from dust and foreign material.
- free from defective, diseased, decayed and damaged bulbs caused by seed stems, tops
- free from moisture, dry sun scald burn, sprouting, mechanical or injuries and staining.
- free from moulds, soft rot and insect attack.

### 7.7.2 Grading of Onion

- Onions after curing are graded manually before they go for storage or marketing.
- Before storage, double, broken, rotten bulbs with undesirable features are removed.
- Grading and classification are required for local and export markets.
- Different markets prefer different sized bulbs, so grading is required by grower.
- The thick neck, bolted, doubles, injured where decayed bulbs are picked out which misshapen small bulbs.
- In this, outer dry scales usually rub off during the process, which gives onions a better appearance for market.
- If storage is arranged after proper sorting and grading, the losses in storage gets reduced.
- For local market, onions are graded based on their size, ie extra large onion (>6 cm dia.), medium (4-6 cm dia.) and small (2-4 cm dia.).
- The extra large onions have great demand and fetches very good price.

### 7.7.3 Sorting/Grading of Garlic Bulbs

- Garlic bulbs after curing are run over grader or can be graded manually before storage or marketing.
- In this, thick- necked, splitted, injured and diseased bulbs with hollow cloves are sorted out.
- Size grading is done after sorting as it is required for better pricing and minimizing losses based on driage and decay.
- It necessary for having better price and to minimize losses based on drying and decay, the government prescribed certain grade designations for different qualities of garlic for export.

## 7.8 Fill the bulb the perforated nylon bags for proper ventilation

### 7.8.1 Packaging of Onion Bulb in bags/nets

- Packing of onion bulb should be small for easy handling during transit and as per market demand.
- Onions are packed in jute/hessian bags for transporting to yard or brought as loose.
- For safe handling, 40 kg open mesh jute bags with 200-300 g weight should be used in domestic market.
- For export, big onions are packed in 5-25 kg size open mesh jute bags.
- Bangalore Rose and multiplier onions are packed for export in 14-15 kg wooden baskets.
- Nylon net bags, when used for packing have resulted in less storage loss due to good ventilation.
- Sacks and nets used for onion packaging for general-purpose jute sacks for agricultural commodities, open-weave sacks of sisal fibre, open-mesh nets of plastic materials and big bags for containing up to 1000 kg.
- Even though, jute sacks are readily available, disadvantages associated with them are large capacity, poor visibility of bulbs through fabric, resistance to airflow, difficulty in labeling and chances of post harvest diseases due to reuse of sacks.
- Sisal sacks are made from sisal-like hard fibres having open weave having thick threads spaced between 10-15 cm apart.
- The rough nature of fibre gives sufficient stable weave and is similar to jute sacks which allows limited visibility of onions and impedance to less airflow.
- Open-mesh nets are used package for onions, which are red or orange in colour.
- The slippery nature of plastics can result in movement of threads allowing large holes to open up.
- To overcome such problem, alternative nets are industrially produced to give fully stable mesh and stronger bag.
- They are degraded by sunlight and should not be left outdoors for long period before use.
- In comparison with other types of bags, they are light weight, small bulk when empty, available in 12.5 and 25 kg sizes, good visibility, excellent ventilation, hygienic, easy closing and crop brand and marketing information may be printed around the middle of bag for easy identification.

### 7.8.2 Storing of Garlic Bulb

- Thoroughly cured garlic bulbs are stored in ordinary well-ventilated rooms.
- Garlic with dried leaves can be stored by hanging in well-ventilated rooms that is not possible on commercial scale due to space requirement.
- Storage without tops in nylon-netted bags give better performance as same has recommended for storage to minimize loss.

- As garlic stores well for market under wide range of temperature, controlled conditioned storage are not necessary.
- Cloves sprout at 4.4°C and prolonged storage in temperature range needs to be avoided, but storage at 0.5°C is good while high humidities often accompany low temperature storage.
- Garlic stored at humidity more than 70% at any temperature develops mould and start rotting.
- Cold storage of garlic is possible at 32-36°F and 60-70% relative humidity.
- The storage loss of 12.5% is recorded in garlic stored at 1-5°C and 75% relative humidity compared to 42.4% losses in ambient temperature.
- Doses more than 10 Kr reduces diallyl disulfide content which gives typical garlic flavour.
- Storage life of garlic is prolonged and loss in eight is also reduced by spraying 3000 ppm MH 3 weeks before harvesting.
- Pre-harvest spraying of 0.1% carbendazim and disinfection of premises for handling and storage of garlic reduces post-harvest losses.

## 7.9 Use of chemical for increasing post harvest treatment shelf life

### 7.9.1 Introduction

- In fields, chemicals are sprayed to control pests and diseases and to increase shell life of crop. When chemicals are used, droplets are produced that can remain suspended in air and gets carried away by wind from target area.

**Fig Using chemicals**

- Chemical spray drift cannot always contained or controlled completely, despite of correct application, so chemical sprays drift over neighbouring properties or waterways and affects human health, animals or environment. This spray drift affects household and farm water supplies, including tank water.
- Chemicals like lawn feed, weed killer and insecticide is what an ailing field needs. There are many kinds of man-made additives that can help in driving pests out from fields and helps in growth and improvement of crop quality. Selecting best chemical for basic needs depends on desired result, hardiness of plants or pests and extent of problem.
- While some chemicals are pre-mixed and ready to apply, certain chemicals can come in concentrated forms. Such stronger chemicals can be mixed by grower in

desired quantities, for treating either minor infestations or more pervasive problems. While using concentrated chemicals, make sure that the instructions written on the packaging should be followed. If a mixture is too strong, the risk is more with killing of your plants. Also, crop growers should always ensure to keep any poisonous items out of the reach of children and pets.

### 7.9.2 Types

Many crop growers choose to use chemicals to keep weeds and pests from destroying their crops and to add more nutrients to the soil. There are three different kinds of pesticides;

- herbicides
- insecticides
- fungicides

### 7.9.3 Chemicals for Improving Shell Life

- Many chemicals are used as dips or dusts to control postharvest diseases of crops/ onions during storage.
- As postharvest diseases start before or after harvesting and infect crops in the field before harvesting, so chemical based agents like chlorine solutions, peroxyacetic acid, organic acids, hydrogen peroxide and electrolysed water needs to be used to make the crops disinfect and to increase shell life.
- Chlorine based solution like NaClO has been commonly used disinfectants for fresh produce, owing to its potent oxidizing properties and cost effectiveness
- Organic acid, ascorbic acid and calcium based solutions can be applied largely to slow down enzymatic and non-enzymatic browning, deterioration of texture and microbial growth on fresh produce.
- Maleic hydrazide has been used to inhibit sprouting of onions during storage.
- Spraying onion with 2500 ppm Maleic hydrazide for 2 weeks before harvest can completely inhibits sprouting during storage.
- Spraying with 908 g of acid equivalent of Maleic hydrazide when 50% of tops are down shows good control of sprouting.
- Sulfur dioxide is widely used on table grapes to prevent decay during storage, by initial fumigation from field followed by weekly fumigation of storage rooms or slow release from in-package pads with sodium metabisulfite
- Ozone is an active agent against spores of bacteria and fungi which leaves no residues on produce as it decomposes rapidly and helps in extending postharvest storage life of crop.
- Onions exposed to ozone had half mold growth as compared to growth on untreated onions after 4 weeks of storage at low temperature.
- Even after 4 weeks of storage, onions stored at 20o C for additional 4 weeks had less than 50% of control.

### 7.9.4 Effect of Chemicals

Effect on Environment

- Chemical pesticides are known to pollute the environment.
- Chemicals have been used on fields across the world for many years, creating a buildup of adverse pollution in environment, which continues to grow with every application.
- When pesticides are applied on surface, they travel outside their intended area of use by air, soil or water.
- It is the way by which, chemical pesticides cause collateral damage, beyond their intended use.

Effects on Soil and Crops

- When crop growers across the world relies on chemical pesticides, a drastic change in soil health is shown.
- When the health of soil is compromised, the nutritional value of food it gives will get compromised.
- These chemical pesticides not only deplete the nutritional value of food, but also contaminate it.

Health Effects

- Pesticides have been linked to a myriad of diseases.
- As per the study, consistent evidence of serious health risks such as cancer, nervous system diseases and reproductive problems in people/worker occurs when they are exposed to pesticides.
- Due to chemical pesticides, here will be certain diseases or disorders which occur such as neurological disorders, Parkinson's disease, childhood leukemia, lymphoma, asthma etc.

## 7.10 Storage of onion in dark and proper ventilated place

- The best way to increase the shelf life and to the keep onions fresh, is to store them whole.
- Keep onions in well ventilated cool dark place at 40-50f temperature as they easily absorb moisture.
- It is important to ensure proper ventilation for storage onion to prevent molding and rotting, so crop grower should use open basket, bamboo steamer, mesh bag, netted bag or pantyhose for storage.

- Onion bulbs need to be stored in 40-50 kg jute bags, gunny bags, plastic netted bags or plastic and wooden baskets for better storage.
- Crop grower should not leave onions in plastic bags, as this will make onion to spoil quickly due to poor ventilation.
- Storing onions in dark place will help onion to last longer as lack of sunlight reduces changes in temperature and humidity, that causes them to go bad faster.
- Since curing will help in removing excess moisture and allow onion to last longer, so best place to store onion is in cool dry, dark and well-ventilated place as such conditions will ensure that onion don't absorb much moisture or experience heat or humidity.

## 7.11 Monitoring the stored onion after every 15 days

- Losses of onion results due to spoilage from microbial infection, sprouting and mass loss.
- Pre-harvest factors like cultivar, climate, growing conditions, harvesting, curing, storage duration, temperature and relative humidity during storage, influence onion losses.
- Onions are stored in bags or in wooden boxes, so monitoring of onion bulb quality while storage is challenging, as bulbs are difficult to access.
- The air at temperature of 68°F with relative humidity of 50%, has dewpoint near 50°F, so if onions at 40°F are exposed to air, moisture will condense on its surface as moisture on surface prevents the removal of loose outer bulb scales, and poses problem in cleaning.
- While storing in bags, net, wooden boxes, diseases may develop without being noticed, as most onion diseases develop inside the bulb, so there is need for regular monitorisation of onion for onion quality/marketing.

## Chapter 8

# Assimilating Market Information

## 8.1 Identification of suitable market for sale of bulb

### 8.1.1 Introduction

Marketing is an important factor in determining the success of bulb crop or related product. It includes all operations and decisions made by producer/grower which can be determination of most marketable crops for production, best delivery quality produce etc. Marketing does not begin after crop is produced, but is an alternative need to be considered even before production takes place.

Marketing of agriculture produce covers assembling, handling, storage, transport, processing, wholesaling, retailing and export of commodities with supporting services like market information, establishment of grades and standards, commodity trade, financing and price risk management.

**Features**

- Many crop growers sells value-added products with high markups online.
- The crop growers are selling dried bulbs dried bulb crops for general use.
- Current market for crops produced in fields include general public, contractors, independent or chain retailers, other production nurseries, brokers, government agencies or a combination of these.
- Production nurseries that sell to general public may do so via an affiliated retail nursery or by mail order.
- Production nurseries selling to retail operations may further specialize in sales to independent retail centers, chain retail centers or big stores.
- Sales to retail centers calls for more diverse group of bulb crop species and higher quality, higher-priced products in greater selection of cultivars and sizes.
- Such market type usually involves many different customers in various locations needing different products at different times.
- A production nursery servicing retail centers has its own trucks for delivering crops.

### 8.1.2 Market Considerations

- One of the first rules of business is to know the potential markets.
- In bulb crop business, one should first determine who is the customer, where its available markets are in relation to its customer base and what products it might produce for those markets.
- It is best to first identify a market and then develop a plan to produce and sell to that market.
- Many crop nurseries are started based on that, the owners personal favorite crop will sell exceptionally well or that a nursery can be a method of getting rid of excess crop materials from personal collections.
- If a market for the product is not first identified and properly exploited, then such nursery will not be so successful in long term.

### 8.1.3 Marketing System of bulb crops

- Marketing channels shows how various market participants are organized to do movement of product from producer to final consumer which are classified as per channel levels where number of channel levels differs by product and by rural and urban location.
- The marketable categories of onion comprises of spring, summer fresh market onions and fall/winter storage onions.
- Fresh market, or green, onions are pulled while the tops are still green and usually before a large bulb has formed.
- Fresh market onions can be identified by thin, light-colored skin and may have edible green tops.
- Fresh market and storage onion markets account for the largest share of onion use, other markets also account for significant production.
- Bulb crop moves from farmers to ultimate consumers through number of marketing channels such as pharia, wholesaler, aratdar and retailer.

| | |
|---|---|
| Channel -1 | Farmers – *pharia* – wholesaler – *aratdar* - retailer - consumer |
| Channel -2 | Farmers – *pharia* - retailer – consumer |
| Channel -3 | Farmers – wholesaler – *aratdar* - retailer – consumer |
| Channel -4 | Farmers - retailer – consumer |

- In this, 55% of produced bulb crop moves from farmers to consumer by Channel- 1 while 22% moves through Channel-3. Also, 10% and 13% of bulb crop moves through Channel-2 and Channel-4.

### 8.1.4 Types of Bulb Crop Market

Market of bulb crops can be classified based on:

Location: On the basis of place of location or operation, bulb crop markets can be:

- ❖ Village Markets
- ❖ Primary wholesale Markets
- ❖ Secondary wholesale Markets
- ❖ Terminal Markets
- ❖ Seaboard Markets

Area/Coverage: On the basis of area from where buyers/sellers approaches for transactions can be classified as:

- ❖ Local or Village Markets
- ❖ Regional Markets
- ❖ National Markets
- ❖ World Market

Time Span:

- ❖ Short-period Markets
- ❖ Long-period Markets
- ❖ Secular Markets

Volume of Transactions: There are two types of markets based on volume of transactions such as:

- ❖ Wholesale Markets
- ❖ Retail Markets

Nature of Transactions: The markets which are based on types of transactions where people are engaged can be:

- ❖ Spot or Cash Markets
- ❖ Forward Markets

Number of Commodities as per Transactions: A market can be general/specialized based on number of commodities where transactions are completed that includes:

- ❖ General Markets
- ❖ Specialized Markets

Degree of Competition: Market can be placed based on continuous scale, right from perfectly competitive to pure monopoly, so on basis of competition, markets can be:

- ❖ Perfect Markets: Where following conditions hold true, such as:
- ❖ Large number of buyers and sellers;
- ❖ Buyers and sellers to have perfect knowledge about demand, supply and prices;
- ❖ Prices at any one time are uniform over geographical area, plus/minus cost of supplies from surplus to deficit areas;
- ❖ Prices are uniform at any one place over periods of time, plus/minus cost of storage from one period to another;

- Prices of different forms of product are uniform, plus/minus cost of converting product from one form to another.

Imperfect Markets: Where conditions of perfect competition are lacking which includes:

- Monopoly Market
- Duopoly Market
- Oligopoly Market
- Monopolistic competition

Nature of Commodities: Based on type of goods dealt in, markets can be:

- Commodity Markets
- Capital Markets

Stage of Marketing: Based on stage of marketing, markets can be:

- Producing Markets
- Consuming Markets

Extent of Public Intervention: Based on extent of public intervention, markets can be:

- Regulated Markets
- Unregulated Markets

Type of Population Served: Based on type of population served, market can be:

- Urban Market
- Rural Market

## 8.2 Collecting crop bulb market information from reliable source

Bulb crop grower should collect the data/information from both primary and secondary sources.

### 8.2.1 Crop Marketing Source

The marketing systems differ widely as per the commodity, systems of production, culture and traditions of producers and level of development of both particular country and particular sector within that country, so crop grower needs to collect market information about bulb crops from:

- marketing boards,
- crop elevators,
- brokers,
- millers,
- farmers,
- animal feed processors,
- food manufacturers,
- exporters

Growers/producers for better information can meet people from direct or non-direct markets and should understand various crop needs, demand, system and pricing.

## Direct Market

- It involves sale of crop by producer to consumer/miller directly, without any middleman.
- It allow crowers and millers and other bulk buyers to economize transportation cost and improve price realization.
- Direct market runs at expenses of state exchequer as promotional measure to encourage marketing by small and marginal farmers.
- In this, big companies enter into contracts with producers for buying their produce.

Direct market involves producer interaction with consumers on one-on-one basis which include:

- pick-your-own operations
- roadside stands
- community supported agriculture
- farmers markets

Benefits of Direct Market

- Increases price realization of the producer.
- Minimizes marketing cost, transportation cost.
- Encourages distributional efficiency.
- Satisfies consumer through better quality of produce at reasonable price.
- Encourage direct interaction between producers and consumers.
- Encourage farmers for retail sale of their produce with involvement in marketing process.
- Help farmers in discovering the demand of markets for future market oriented planning.

## Non-direct markets

It involve producer interaction with market intermediaries which include:

- terminal market firms
- shipping point firms
- processors
- grower cooperatives
- brokers
- retail outlets

## 8.3 Taking right decision at right time for sale of crop bulb

### 8.3.1 Marketing Cost of bulb crop

For sale of bulb crop, crop grower should know about:

- marketing cost of crop which describes about the cost that involves marketing functions which are required to transfer commodity from place of production to ultimate consumers which includes cost of transport, loading and unloading, gunny bag, commission, market tolls, bribery/subscription, storage, house fair, cleaning and grading, and others.
- type and location of customers that greatly affect marketing costs as many retail sales needs colour point of sale tags, comprehensive marketing plan, and branding program to maximize exposure, impact and sales to retail customer.
- marketing costs link with sales to landscapers and other nurseries which are not burdened by marketing expenses, as there are costs associated with trade shows, websites/catalogs and plant locator services.
- nursery sales which may be handled by in-house sales staff, brokers or combination of both.

### 8.3.2 Market Price

Movement of products from producers to consumers involves costs, taxes and cess which together contribute to market price. Such price vary with channels by which a particular crop/commodity passes through such as cost of packing, transport, weighment, loading, unloading, losses and spoilages.

Market price normally includes:

- Handling charges at local point
- Assembling charges
- Transportation and storage charges
- Handling by wholesaler and retailer charges to customers
- Expenses on secondary service like financing, risk taking and market intelligence
- Profit margins taken out by different agencies.
- Producer's share in consumer's rupee

### 8.3.3 Market margins and price

For decision of bulb crop pricing, crop grower should know about:

- price spread which shows difference between price paid by consumer and price received by producer for certain quantity of crop produce.
- what price spread comprises of such as marketing costs and margins of intermediaries to determine overall effectiveness of marketing system.
- were highest margins are there for retailer in comparison with intermediators in secondary and terminal markets.

- were highest profits are there for retailer in secondary and terminal market.
- profit as percent of investment was maximum for retailer.
- reason for highest profit and margins due to high selling price of bulb crops.
- market margin of crop as calculated by deducting purchase price from sale price and profit from difference between the marketing margin and marketing cost.

### 8.3.4 Decision based on Price Fluctuation

- Bulb crop growers/farmers mostly have downside price risk in contrast to end user who mostly has upside price risk.
- Managing price risk can occur through a spot transaction if the product is available or by forward transaction if crop/product is not available, either with physical forward contract or hedging price by futures market.
- Selecting subjective price targets/triggers depends on individual risk attitude and circumstance of decision maker.
- An unrealistic target price adds profit margin to current price in bearish market when commodity market intrinsically ignores operational costs of farmer or end user.
- Crop growers sometime are forced to accept prices below their cost of production, whereas end users may sometimes be forced to buy above their break-even limit.
- The alternative for farmers is to withhold product from the supply chain either by hoarding or pooling, whereas the alternative for end users is a switch to substitute products or alternate operating methods.
- To minimize price risk, farmers/crop growers should understand the meaning of Minimum Support Price that is a form of market intervention which insure agricultural producers against any sharp fall in prices.
- The minimum support prices are announced by government at start of crop sowing season on the basis of recommendations of Commission for Agricultural Costs and Prices.
- Minimum support prices are guarantee price for crop produce from government as its main objectives is to support the farmers from distress sales and to procure food grains for public distribution.
- In case market price for commodity falls below the announced minimum price due to bumper production and glut in the market, government agencies purchase the entire quantity offered by the farmers at the announced minimum price.

## 8.4 Analyse previous year rate of bulb crop

### 8.4.1 Information

- Market information provides hints to crop growers on overall performance of crop through price signals and linking food deficit areas to surplus areas, domestic markets to international markets and vice versa.
- The study of marketing costs and margins about bulb crops serves in evaluation of marketing performance for input and output markets and across the entire market or value chain.

- Farmers usually assess prices before selling bulb crop in the market.
- By tracing market cost of crop, an estimate of magnitude and quantity traded can be analysed from official records.
- Results from survey shows that brokers are main information source, which accounts 85% for farmers followed by 9%, which is information obtained from other farmers.
- Farmers get market information from brokers who often have the power to set the price to farmers.
- Due to seasonality, nature of production in district, smallholder farmers are always uncertain, which attributes to rise and fall of price in market.
- During a shortage of bulb crop in the market selling price rises while during peak production of bulb crop, price falls on average due to excess supply and absence of linking crop to export market destination.

### 8.4.2 Price volatility

- Crop price volatility remained a major concern despite considerable advances in communication, transportation and production technologies as crop production is closely associated with national food security and welfare of farming population, so managing price instability has been a major challenge. Crop producers want prices to be stabilized at levels that guarantee adequate returns but interventions to achieve such objective are often costly to consumers and tax payers.

The volatility is often attributed to special characteristics of agriculture which includes:

- supply disturbances (climate and pests)
- inelastic supply and demand
- production decisions based on current prices
- fluctuating food
- agricultural policies

Price volatility that results from peaks and troughs in production is often made worse by host of endogenous causes such as:

- lack of adequate storage facilities
- low stock levels
- sudden jumps in input prices
- underdeveloped market infrastructure
- limited export and import
- High levels of production and low prices in agricultural sector also follow shortages and high prices in unstable cobweb process if production decisions at planting time are based on expectation that current prices will remain in place after harvest. If current prices are high, wheat growers/producers assume that this situation will continue and will accordingly take measures to expand production for next season, which results in high levels of production and price collapse.

## 8.5 Close monitoring of various platform for rate update: SMS, Newspaper/TV/Radio

### 8.5.1 Introduction

- Agriculture information should reach to farmers in right perspective, in right form and in right time as it is very critical for enhancing the productivity of farm sector.

**Fig Getting information on telephone**

- Farmers are getting agricultural information from diverse communication sources like neighbours, local leaders, extension agents, newspapers, telephone, farm magazines, television, radio etc. In context of rural development, nothing is more important than the transfer of useful ideas from one person to another.

### 8.5.2 Sources of information

As information on agriculture in terms of crop and livestock were communicated among farmers from ancient times, so with development in agricultural research, the need arises to rapidly and accurately transfer new information and technologies to farmers.

To fulfill such need, media such as print and electronic and interpersonal communication has proved its usefulness in creating awareness about new agricultural technologies.

1. Interpersonal sources

It includes:

- personal localite
- extension agents
- friends
- family members
- fellow farmers
- neighbours
- group discussions
- demonstrations
- campaign
- village leaders

The sources play significant role in dissemination of information related to agriculture. During pre-independence, interpersonal communication channels were used widely. Demonstration centers were established to disseminate information by conducted demonstration under guidance of experts.

2. Mass media

❖ Mass media serves as channels of communication that expose large numbers of people to similar information at same time.

❖ It includes:

  - media that conveys information by sound such as radio, audio cassettes;
  - moving pictures such as television, film, video;
  - print such as posters, newspapers, leaflets

❖ The attraction of mass media to extension services is its high speed with low cost through which information can be communicated to people over wide area.

❖ As cost of producing and transmitting a radio programme may seem high, when that cost is divided between millions of people who may hear programme, it results as cheaper way of information.

❖ The cost of an hour's radio broadcast per farmer who listens can be less than one-hundredth of cost of an hour's contact with an extension agent.

3. Radio

❖ It is a popular medium for information as well as attitude change.

❖ Radio from its inception played an important role in development communication.

❖ It plays a peculiar role in technology dissemination.

❖ It is very effective media which can reach effectively to rural people.

❖ In India, All India Radio has been the forerunner in process of implementing communication strategy adopted by government.

❖ Radio communication works roughly in 15000 villages and carries certain programs like Utam Khetti, Kisan Vani, Navi Chetana on agriculture.

4. Television

❖ It is an ideal medium to convey information and news to illiterate and literate, urban and rural viewers on whom it would have profound impact.

❖ It is used in planned manner to motivate people to participate in developmental programs.

❖ It is introduced in India in 1959 on experimental basis.

❖ In 1967, Delhi Television centre launched Krishi Darshan Program which gained popularity as they have been able to solve numerous ground level problems of farmers and also replaced their antiquated technique with new innovations.

### 8.5.3 Effectiveness of mass media/interpersonal communication by crop growers

❖ As studied on farmer's information source and their relative effectiveness, it is noticed that print media were regarded as most effective source, as they were used as source of information by only 11% of respondents.

- The information sources for agricultural technology found that contact farmers were accorded first rank by respondents on credibility.
- The role of mass media in dissemination of agricultural technologies among farmers revealed that T.V was ranked 1$^{st}$ with regard to its effectiveness in dissemination of agricultural technologies while newspapers, posters, charts, books/booklets and bulletins ranked 3$^{rd}$, 4$^{th}$, 5$^{th}$, 6$^{th}$ and 7$^{th}$ on basis of effectiveness.
- With role of television in agriculture development, majority of respondents felt that television is indeed the best medium of disseminating agricultural related information to general public.
- Many farmers found that farm magazine gives highest satisfaction followed by telephone, television, computer and radio.

# Chapter 9

# Basic Farm Management

## 9.1 Calculate cost of production of bulb crop

Onion is most popular bulb crop which is much higher in rabbi season as compared to in Kharif season. The cost of cultivation of onion crop per hectare depends on following factors:

### Variable Cost

It includes:

A) Soil fumigation

B) Land Preparation: It covers information such as:

- Bullock /Tractor- 3 Ploughings = 3 Plankings
- Bed and channel preparation
- Transplanting and mulching
- Interculture
- Irrigation at seedling level and crop growth
- Fertilizers and manures
- Plant protection
- Harvesting
- Transportation

C) Seed Cost

D) Fertilizers and manures

E) Pesticides/plant protection

F) Miscellaneous (2% of working Capital) =

G) Interest on working capital (6% of working capital)

### Fixed cost

It includes:

A) Land revenue (52 paisa/bigha)

B) Rental value of land (1000/bigha)

C) Management cost (10% of working capital)

D) Risk margin (10% of working capital)

E) Depreciation cost (Rs 200/bigha)

F) Interest on fixed capital (6% of working capital)

To calculate cost of cultivation of bulb crop, crop grower should use following formula:

Total cost = Total variable cost + Total fixed cost

where:

- Total cost of cultivation = Total variable cost + Total fixed cost
- Total income = Yield (kg) × Market price of the crop (Rs./kg)
- Net Profit = Total Income - Total cost of cultivation
- Benefit cost Ratio = Cost of total benefit / Cost of production

## 9.2 Soil testing, selection of crop variety, time of sowing, crop rotation, intercropping, schedule for fertilizer, pesticide/chemical application, irrigation scheduling and harvesting etc.

### 9.2.1 Soil testing

Soil testing has been used to formulate precise recommendations for major nutrients i.e. nitrogen, phosphorus fertilization of crops in different soils and to recommend appropriate doses of amendments for salt-affected and acidic soils.

Micronutrients, comprising zinc, copper, iron manganese, boron and chlorine, though required by plants in much smaller amounts, yet are as essential for them as the major nutrients. Despite that, little attention has been paid to employ the soil testing for assessing the micronutrient status of soils and determining soils requirement for micronutrient fertilizers for growing crops. For proper determination of soil nutrients, soil should be tested at the laboratories.

#### Testing in laboratory

- Testing of soil is normally done to assess its physical properties which affect or control its behavior in relation to various purposes.
- Although testing activities include on-site/in-situ and laboratory testing, the latter one is assumed to give a better and accurate estimate of its properties.
- Each of the different properties of soil gives an estimate of a particular characteristic of it which includes moisture content of a soil specimen to guide the compaction process, besides governing the Atterberg limits, consistency, shrinkage characteristics, and a host of other strength properties.
- Size-grading of soil helps in the optimum designing for preparing concrete mixes and porosity along-with permeability tests are conducted on a soil sample to determine the suitability of semi-pervious materials for use in the core of fill dams.

- The process of soil testing requires mixing several random soil samples before submitting for testing.
- Results of the test give information to assist the homeowner in developing a fertilization program.
- Each soil testing laboratory uses different testing techniques, it is critical that the lab provide an interpretation of the soil test as a recommendation for applying or withholding fertilizers.
- The choice of plants to add to a landscape can also be directed in part through the information produced in the test.

## Soil Sample Selection

- The testing of soil in the laboratory requires only few grams of soil sample.
- In a homogenous field, soil samples from plough layer (0-15cm) should be selected randomly in a zig-zag manner. The samples should not be collected from near the bunds, water channels, field paths and heaps of crop straw, stubbles, manure, etc.
- The sample collected from the selected sites should be composite and mixed thoroughly in a container.
- From this lot a representative sample, about 500 gm should be taken out and air-dried under shade.
- The air-dried sample should be transferred into a clean cloth bag bearing a slip with a mention of complete address, field number, cropping sequence being followed, source of irrigation, soil type, fertilizer/manure schedule followed in the preceding crops and any other specific observation about the soil and/or the crops grown therein.
- Then the sample should be taken to the laboratory where facilities for testing soils for micronutrients are available.

## Soil Test Report

- Many laboratories involves in standard soil test which covers set of soil chemical measurements that are important in agronomic decision making.
- Certain labs vary, typically which includes pH, buffer pH, cation exchange capacity (CEC), base saturation, organic matter (OM), and the macronutrients phosphorus (P), potassium ($K^+$), calcium ($Ca^{2+}$), and magnesium ($Mg^{2+}$).
- In plains and western regions where salinity and sodicity are more common, conductivity, soluble salts, sodium ($Na^+$) and chloride ($Cl^-$) may also be reported in a local laboratory's standard test.
- Nitrate ($NO^{3-}$) and sulfur (S), are not usually part of standard tests, but are considered macronutrients.
- Nitrate tests are used in some regions for preplant nitrogen (N) management and in-season sidedress decisions, but recommendations vary from region to region.

## Soil Test Values

### pH

- ❖ It is a measure of acidity of a substance.
- ❖ It is expressed as a number, not a percentage or other quantity.
- ❖ A value of 7.0 on pH scale is considered neutral, the substance is neither acidic nor basic.
- ❖ Lower numbers on the pH scale denote increasing $H^+$ ions and acidity.
- ❖ The acidity of the soil solution affects many physical, chemical, and biological reactions necessary for plants to survive and thrive.
- ❖ The optimum pH value varies by crop and region, but is generally between 6.0 and 7.0 pH (slightly acidic), though alfalfa thrives between pH 6.8 to 7.0.

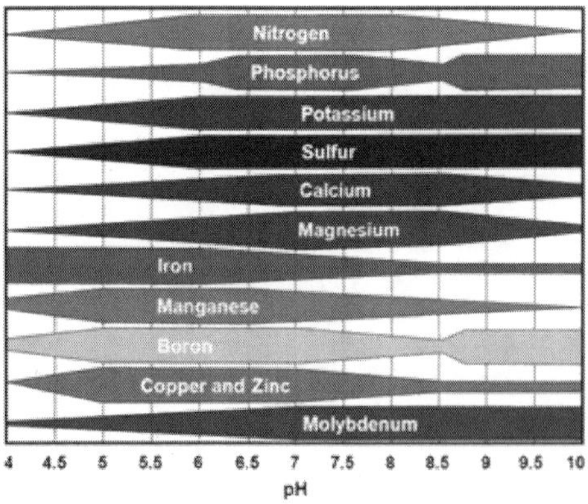


### Buffer pH

- ❖ Once need for adjusting pH higher is identified, buffer pH value shows the amount of liming material required to make the adjustment.
- ❖ Low buffer pH values shows that more lime materials are necessary to raise pH than higher buffer pH values.
- ❖ Buffering refers to ability of the soil to recharge acidity, and is a function of aluminum ($Al^{3+}$) minerals and $H^+$ in the soil.
- ❖ If a soil is poorly buffered (sandy), less liming material are needed to raise pH a certain level than a soil that is highly buffered (clayey soil).
- ❖ Some soil tests may report buffer pH as **lime test index**, or **LTI**, which is merely the buffer pH value multiplied by 10.

## 9.2.2 Identification of appropriate variety of bulb crop

Bulb crops consist of:

- ❖ garlic
- ❖ leeks
- ❖ dry-bulb
- ❖ green onions

## Garlic bulb

- Garlic is hardy perennial member of onion family.
- Garlic is native to Central Asia but has long been naturalized in southern Europe and throughout the world.
- It differs from onion, producing number of small bulbs called as cloves rather than one large bulb.
- In this, each bulb has dozen or more cloves that are covered with thin white skin.
- Each clove is made of two modified mature leaves around an axis with vegetative growing point.
- The outer leaf is dry sheath, while base of inner leaf is thickened, making up the bulk of the clove.
- In this, larger outer clove produces best garlic.
- Garlic has flat leaves rather than round hollow leaves of onion.
- It is used as condiment and flavoring in gravies, tomato sauces, soups, stews, pickles, salads, salad dressing and breads.

## Leek

- It is a non-bulb forming member of onion family which is grown for its blanched stem and leaves.
- It is eaten raw alone or mixed in salads and cooked as flavouring in soups and stews.
- It requires similar climatic and soil requirements as that of onion.
- Except of its large size, it looks like green onion.
- It is cultivated for blanched stems and leaves.
- It is biennial plant, consumed as salad and used in flavouring the soups.
- It is produced on 50 to 100 acres in small tracts for roadside and local grocery sales.

**Onion bulb**

- Onion is an herbaceous biennial in the family Liliaceae grown for its edible bulb.
- The stem of the plant is flattened disc at base and tubular leaves form pseudostem where their sheaths overlap.
- The leaves are either erect or oblique and there are 3–8 per plant.
- The onion plant produces pink or white flowers clustered on stalks.
- The bulbs are formed just above the flattened stem of the plant by overlapping leaves.
- The bulb is made of many layers, each corresponds to leaf.
- They are oval but shape can be variable and occur in clusters of 3–18 to a plant.
- The bulb is protected by a membrane which turns to a papery coat.
- Onion plants can reach a height of 50 cm and are grown as annuals, harvested after one growing season.
- Onion is referred to by cultivar which includes red or purple, shallots and spring or scallions.
- The bulb is edible vegetable and is used part of onion, that consumes after cooking although it can be eaten fresh.
- The stems and leaves are also edible.
- Onion crop can be raised by transplanting, planting bulbs directly in field, broadcasting or dibbling the seeds in prepared field.

**Select appropriate bulb crop variety as per maximum yield**

Yield parameters of bulb crop includes:

- harvest index
- total bulb yield
- marketable bulb yield
- unmarketable bulb yield
- total bulb number
- marketable bulb number
- unmarketable bulb number
- bulb quality parameters

Bulb Quality Parameters include:

- Small Size Bulb Number: The small size bulbs number was obtained per plot at harvest by counting the number of bulbs/plot which was converted to ton/dunam.
- Medium Size Bulb Number: The medium size bulbs number was obtained per plot at harvest by counting number of bulbs/plot which was converted to ton/dunam.
- Large Size Bulb Number: The large size bulbs number was obtained per plot at harvest by counting number of bulbs/plot which was converted to ton/dunam.

## Select bulb crop variety as per resistant to insect, pest and disease

- Insect pests are responsible for substantial crop losses by direct damage and transmission of plant diseases.
- There are many strategies that improves crop resistance to insect pests which are related to plant secondary metabolism and immunity and microbiome science.
- Crop resistance to insects is cultural control methods which involves use of agronomic practices to reduce insect pest abundance and in damaging which occurs if there involves no practice.
- In integrated pest management, plant resistance to insects uses resistant crop varieties to suppress insect pest damage where plant resistance is used in conjunction with other direct control tactics.
- Crop resistance results as consequence of heritable plant qualities which result in plant that is being relatively less damaged as compared to plant without the qualities.
- The insect resistant crop cultivar yields more than susceptible cultivar when confronted with insect pest invasion.
- Resistance of plants is relative and is based on comparison with plants lacking the resistance characters such as susceptible plants.
- Insect resistant crop varieties suppress insect pest abundance or elevate the damage tolerance level of crop.
- The insect resistant plants alter the relationship which an insect pest has with its plant host.
- The use of insect resistant crop varieties is economically, ecologically and environmentally advantageous where economic benefits takes place due to crop yields that saves from loss to insect pests and money by not applying insecticides that would have been applied to susceptible varieties.
- The seed of insect resistant cultivars costs no more, or little more than for susceptible cultivars as ecological and environmental benefits arises from increases in species diversity in agroecosystem due to reduced use of insecticides.
- Increases in species diversity increases ecosystem stability which promotes more sustainable system far less polluted and detrimental to natural resources.
- The main advantage of using insect resistant crop varieties as a component of pest management arises from ecological compatibility and compatibility with other direct control tactics.
- Plant resistance to insect pests has advantages over other direct control tactics as plant resistance to insects is compatible with insecticide use, while biological control is not.

## Select bulb crop variety as per draught tolerant

Drought tolerant or drought resistant bulb crops are those that can withstand prolonged periods of dry weather and often have succulent leaves or stems and can sometimes store water. Other crops have fine hairs or silvery colouring of foliage that cuts down evaporation.

There are many suitable crops/plants that will cope in full sun and look good in a gravel setting having drought tolerant characteristics as silver or grey-green leaves to reflect sun and coating of fine hairs on leaves or stems, for trapping in moisture.

Bulbs are an excellent choice in the garden of the future, and with their natural moisture storing structure, most bulbs tend to be more drought resistant than plants without this adaptation.

And while many consumers are cancelling or postponing their landscaping plans over worries of water shortages and drought-raised irrigation costs, savvy homeowners and landscapers see the drought as an opportunity to replace the often mostly non-native, water-thirsty plants with more efficient gardens, using bulbs and succulents.

Bulbs are an excellent choice in the garden of the future, and with their natural moisture storing structure, most bulbs tend to be more drought resistant than plants without this adaptation. Many of these types of bulbs tolerate drought conditions, especially in the summer while dormant. Storing their entire life cycle in a swollen underground stem, the bulb contains the leaves, flowers, food and the bulblet.

While there are a number of drought tolerant vegetables varieties available, without some planning, extreme drought and heat will kill even the hardiest. Planting at the correct time is crucial. Sow the seeds earlier in the spring to take advantage of the warm weather and jump start the growing season, or plant later in the fall to minimize the use of irrigation and use the seasonal rains to your advantage.

Garlic and perennial onions are alliums with similar growing and storage requirements. Under the right conditions these are very productive crops, with harvests 3–8 times greater by weight than the amount planted. With the right care these alliums are self-sustaining, so they are particularly appropriate if you are interested in local food, permaculture and perennial crops.

It is essential that the soil be kept moist during the entire period of active. If the soil is allowed to dry out just once, growth will be arrested, resulting in smaller bulbs and reduced yields. Heavy mulch between rows and plants will conserve moisture and control weeds.

When planted at the proper depth, bulb clusters of multiplier onions form near the soil surface, not beneath the soil. This allows the bulbs to become larger, reduces susceptibility to rot, and improves keeping quality. Bulbs of garlic and topset onions develop from cloves beneath the soil surface.

The onion plant was shown to be tolerant to water stress, being able to grow and produce seed yields in arid zones with limited amounts of rainfall and without irrigation. Under these conditions, large mother bulbs and high plant density gave the best results. Nitrogen was very important for the production of seed yield, but high amounts applied before planting and additional supply during the growing-season as top dressing did not have a significant effect on seed yield. Similarly, enrichment of the soil with phosphorus or potassium did not increase yield.

### 9.2.3 Time of sowing

| Seasons | Time of sowing |
|---|---|
| In state of **Maharashtra and parts of Gujrat,** sowing time of bulb crop is: | |
| 1. Early Kharif<br>2. Kharif<br>3. Rabi | May-June<br>August-September<br>October -November |
| In state of **Tamil Nadu/Karnataka/A.P.** sowing time of bulb crop is: | |
| 1. Early Kharif<br>2. Kharif<br>3. Rabi | March-April<br>May-June<br>September-October |
| In state of **Rajasthan/Haryana/Punjab/U.P, Bihar** sowing time of bulb crop is: | |
| 1. Kharif<br>2. Rabi | May-June<br>October -November |
| In state of **West Bengal and Orissa,** sowing time of bulb crop is: | |
| 1. Kharif<br>2. Late Kharif | June- July<br>August-September |
| **On hills,** sowing time of bulb crop is: | |
| 1. Rabi<br>2. Summer | September-October<br>November-December |

### 9.2.4 Crop rotation

- Crop rotation grows different crops in same area of land every year.
- Different plant species needs different nutrients from soil and also replenish it with different nutrients, so well-planned crop rotation system will help in combating soil deficiencies and helps in improving mineral and organic content.
- It helps in eradicating weeds, insects and other pests which tend to attack specific plant families or species.
- Rotating crops planted in particular site will break the cycles of parasites which reduces number of potential attackers.
- As onion is self-tolerant, so too narrow crop rotations will lead to development of diseases with origin in soil that are difficult to combat, so a distance of 4–5 years in crop rotation needs to be followed.

If in a field, there are three segments, the crop rotation for leek, garlic and onion for every year for three year rotation plan should be:

**Year one**

Section one: leek

Section two: onions and roots

Section three: garlic

**Year two**

Section one: onions and roots

Section two: garlics

Section three: leek

**Year three**

Section one: garlic

Section two: leek

Section three: onions and roots

### 9.2.5 Schedule for fertilizer

**Introduction**

- ❖ Onions are heavy fertilizer feeders with shallow root systems, so proper soil fertilization is required for obtaining high yield.
- ❖ Proper soil fertility is important in areas where onion makes rapid growth before bulb formation gets triggered by decreasing day lengths.
- ❖ The crop requires different nutrient rates and ratios at different growth stages and in order for the nutrients to be available when the plant needs them, fertilizers should be applied at the right timing.
- ❖ Nitrogen is important for preparing soil for bulbs to be able to start their leaf growth. Phosphorus and potash are good for overall health, resisting disease, root growth and flowering.

**Time of Usage**

Before Planting:

- ❖ Before planting onion transplants or sets, amend the soil with 3 inches of compost to improve soil texture and for slow-release of nutrients.
- ❖ Add 1/2 cup balanced fertilizer per 10 feet of row till amendments to depth of 8 inches.
- ❖ Use 1/2 cup super phosphate fertilizer instead of balanced fertilizer if planting onion seed.
- ❖ Use super phosphate as it acts as starter fertilizer which encourages strong, early roots.

Growing Season

- ❖ Onions need steady supply of nitrogen to form large bulbs.
- ❖ Side dress growing plants in early and mid summer with 1/2 cup nitrogen-based fertilizer.
- ❖ Use ammonium sulfate if the soil is alkaline as such fertilizer lowers the pH.
- ❖ Use ammonium nitrate in acidic soil for good growth.

Precautions

- Avoid composts that contain biosolids or sludge in the vegetable garden.
- Do not use fertilizer after midsummer as late applications of fertilizer encourages new, soft growth which won't keep onion well.
- In addition to adequate fertilizer, onions need evenly moist soil to grow large.
- Use granular fertilizers which has tiny particles which don't dissolve quickly in water and remains in soil for long time.
- Use fertilizer that's higher in phosphorus and potash and lowers in nitrogen, and apply it at half-strength when bulb leaves are out of ground.
- The proportioning and the control of the nitrogen supply is especially important. In practice usually too much nitrogen is applied rather than too few.

### 9.2.6 Pesticide/chemical application

- Pesticides are important tools in managing crop pests, but their applications in enclosed spaces increase the risk of exposure.
- Potential benefits of application of pesticides lead to increased in yield or quality and to outweigh hazard and expense of pesticide application.
- Selecting optimum pesticide and application method depends on pest, equipments, harvest intervals, legality of pesticide on specific crop or plant variety and safety.
- Proper technique of application of pesticide and equipment used for applying pesticide are main for success of pest control operations.
- Application of pesticide is not simply operation of sprayer or duster, but also involves thorough knowledge about pest and its related problems.
- The mode of action of pesticide, its relative toxicity and physicochemical properties, helps in handling precautions, agitation requirement etc.
- Pesticide application besides keeping the pest population under control, also avoids pollution and damage to non targets.
- In this, different types of pesticides are used for controlling various pests, ie insecticides are applied against insect pests, fungicides against crop diseases while herbicides are applied against weeds to protect crop losses.
- Pesticides are dispersed by different methods like spraying, dusting etc.

### 9.2.7 Irrigation scheduling

#### Introduction

- Irrigation scheduling is directly related to profitable crop production and sustainable agricultural practices.
- The frequency of irrigation of farms was determined by prevailing weather conditions without considering the growth stage of the crop.
- Onion is a shallow rooted crop which requires frequent irrigation at short interval.

- ❖ Adequate water is required during initial plant establishment and at bulb development through maturity to produce good yield.
- ❖ Withholding irrigation at 3- and 7-leaf stages reduced total yield of bulb crop as compared with the control.
- ❖ Crop growers who irrigate their farms at 2-3 days per week highlights that irrigation done every other day in prolonged dry periods and less frequently in cloudy weather conditions to prevent high disease incidences.
- ❖ In onion, regardless to type of irrigation system used, yield and quality suffers if irrigation is delayed and available soil moisture is allowed to drop too low.
- ❖ In order for irrigation schedule to be effective, it must tell the grower when and how much water should be applied.

**Importance**

- ❖ Onions have little tolerance for water stress.
- ❖ Bulb market grade and single centeredness are influenced by small increments of water stress.
- ❖ The incentives to maintain precise irrigation schedule to keep soil water potential in narrow range of values are significant.
- ❖ Optimal irrigation scheduling leads to maximum yields and good bulb quality.
- ❖ Under-irrigation leads to losses in total yield, market grade, and single centeredness.
- ❖ Over-irrigation leads to soil erosion, bulb disease susceptibility, water loss, extra energy costs for pumping, N leaching, and increased crop N needs.

**Soil water tension method:**

- ❖ Onion plant performance is related to amount of tension the plant has to exert to move water from soil into plant roots.
- ❖ This method of irrigation scheduling measure, how strongly water is held by the soil.
- ❖ The higher the soil water tension, the drier the soil.
- ❖ The method consists of establishing a maximum soil water tension as criterion for beginning irrigation and to irrigate when the method is reached.
- ❖ This method can be measured using tensiometers, Granular Matrix Sensors (GMS), or other instruments.
- ❖ Granular Matrix Sensors has two electrodes in porous matrix where electrical resistance between two electrodes is measured and converted to soil water tension using equations.
- ❖ In this, six or more Granular Matrix Sensors can characterize soil water tension in a field, provided they are installed in representative areas responsive to ET and irrigations.
- ❖ In this, sensors are kept below double rows of onions between three to four healthy plants where placement of soil moisture sensors at a depth central to crop root zone is important.

- The GMS can be read using hand-held battery-powered meter or by dataloggers.

**Crop Evapotranspiration (ET) method**

- It is the combined evaporation of water from soil surface and crop water use.
- In this crop evapotranspiration values can be calculated using weather stations in production region.
- In this, long-day onions grown in region uses much less water per day at beginning and end of season as compared to middle of season.
- Daily onion ET ranges from about 0.05 inch at beginning of season to 0.5 inch in middle of season.
- Here total seasonal onion evapotranspiration is in the range of 28-32 inches of water.

## 9.3 Maintain record of investment and expenditure

### 9.3.1 Records

Keeping track of what is happening on the farm requires some records. Good farm management requires good useful set of farm records which does not ensure successful of farms, but success is unlikely without them.

Record-keeping is a task that has been started many times by eager farmers, only to be dropped later. Reasons that complete, accurate records are not kept revolve around lack of time and motivation. Keeping records is productive work.

Good record-keeping system should take into account the wheat growers needs, abilities and limitations. Wheat growers should need to understand the purpose for keeping the records and when the purpose is known, it will help them in deciding which records are useful and need to be kept.

With accurate facts and figures is useful while borrowing money, seeking government support and completing tax returns. From reliable farm records, equity proportion of total assets owned can be updated to assist with future farm investment programs.

Farm records can help wheat growers in evaluating wheat enterprise on mixed farm while farm advice will be more effective if adviser knows exactly what is happening on the farm. Records and statements show what has occurred or is occurring on the farm while budgets predict what might happen in the future.

Records provide wheat grower with data, information and knowledge. There are four uses for farm records such as:

- Use as service tool which covers income tax calculations, estate planning, business arrangement reconciliation and obtaining and managing credit.
- Providing data for financial analysis and diagnostic instruments like identifying strengths and weaknesses of business.
- Progress indicator where series of records are required to monitor work progress.
- Forward planning tool for keeping past records to project future cash flows for current and modified farm management practices.

### 9.3.2 Details of Record

Records of farm comprises of:

Farm diary

- It is a form of small notebook where wheat grower should record key facts and figures about farm business and its day-to-day activities.
- Farmers /wheat grower who use diary finds important facts and figures which could be easily lost or forgotten are stored permanently in this for future reference.

Crop records

It is useful to record what happens to each plot or crop each year that will includes:

- crop,
- fertilizer applications,
- agronomy (seeding, weeding, sprays),
- visual impression of crop,
- harvest interval of forage crops
- crop yield

Unit costs

It covers all main farm inputs like:

- fertilizers,
- fuel,
- irrigation water,
- concentrates and their ingredients,
- purchased forages,
- stock purchases

The details are required for routine book keeping and in monitoring seasonal changes and will help in planning future purchases.

Unit returns

- It includes returns from all farm outputs such as sale of crop, manure etc.

Crop stock accounting

- The main purpose of crop stock accounts is to monitor net losses and gains in income, and to distinguish the increases and decreases due to changes in market value.
- When quantifying total stock value, wheat grower has to take into account both in change size and change in unit of the crop.

Plant and improvement records

It covers records related to:

- date and cost of purchase or installation,
- annual depreciation,

- insurance and registration,
- fuel use,
- hours use,
- repairs and maintenance

Key financial records

- It covers detail related to interest and principal repayment schedules to plan repayments.

Other Farm Charges

It include records related to:

- rates and government charges,
- telephone and office upkeep charges,
- labourer's wages,
- capital investments,
- unusual weather events,
- important meetings,
- credit repayments,
- changes in valuation of capital

Personal expenses

- It includes expenses related to personal input wages which includes valuation of any farm produce used for personal consumption.

Different records

- It covers cost related to crops, fertilizer, seeds, spray, labour, water, machinery, harvesting, processing, transport and selling.
- In this, wages are broken down into different farm activities.

### 9.3.3 Importance of Records

- They are management tool which allows wheat growers to measure how efficiently they are using resources and to determine whether or not they are having income.
- The records are essential for planning and decision making
- The records are used in obtaining credit
- The records are used for income tax

### 9.3.4 Farm Accounting

It is the process of measuring and recording all farm resources and all farm transactions with financial consequences in a systematic way. The conduct of farm accounting takes large part of wheat grower/farmer's time and effort which gets compensated in account of following reasons:

- ❖ It permits farmer/wheat grower to find out the size of income which is derived from the farm.
- ❖ It helps in knowing the total value of farm business and to know which part is actually owned by him/her and which by others.
- ❖ It will help in detecting loss or theft of cash or stock.
- ❖ It provides necessary data for correct income tax assessment.
- ❖ It helps in claiming expenses.
- ❖ It helps in keeping complete information as farmers/wheat grower dislike paper work since they are busy with their farm work.

### 9.3.5 Farm Financial Records

- ❖ Farm transactions can be recorded when they occur or at the time of cash transactions. Since farmers are involved in lot of cash transition, so they should use accrual method of accounting, which will help them in tracking of their money, inventory, sales etc. For example, when recording sales of wheat for specific time period, all the expenses needed to generate that wheat are relevant when calculating the cost of production.

Every physical item on farm can be placed into following categories:

1. Assets: It is required to generate income which can be either owned/purchased on credit that can be combination of physical and monetary values, where physical quantity is multiplied by unit price to have its value.
2. Liabilities: It relates to items on credit.
3. Equity: It serves as an asset with less liabilities.
4. Revenue: It is source of income that is generated from the business.
5. Expenses: It involves the costs what is required for a business to generate revenue.

Annual financial accounting comprises of:

1. Opening balance sheet: record all assets and liabilities at beginning of financial period.
2. Closing balance sheet: record all assets and liabilities at end of financial period.
3. Cash flow: record all cash inflows and outflows in business.
4. Income statement: calculate wheat enterprise profit/loss before or after income tax.

## 9.5 Identification of near market for sale of bulb

Crop Growers for better selling of his/her crop should meet people from direct or non-direct markets and should understand the scope of various crop needs, demand, system and pricing.

### Direct Market

The direct market involves selling of crop through interaction with:

- pick-your-own operations
- roadside stands
- community supported agriculture
- farmers markets

### Non-direct markets

It involve selling of crop with:

- terminal market firms
- cooperatives
- brokers
- retail outlets

# References


Amha, W., 1994. Food Grain Marketing Development in Ethiopia after the Market Reform 1990: A Case Study of Alaba Siraro District. Koster Publ., Germany, ISBN-13: 9783895740060, Pages: 292.

Anonymous, 2005. Book of extended abstracts (General Biology, early life history and conservation-trade). Proceedings of the 5th International Symposium on Sturgeon, May 9-13, 2005, Ramsar, Iran, pp: 151-328.

Bahta, S.T. and S. Bauer, 2007. Analysis of the determinants of market participation within the South African small-scale livestock sector. Proceedings of the Conference on International Agricultural Research for Development, October 9-11, 2007, Witzenhausen, Germany -.

Bedada, D.A., 2003. Analysis of onion marketing: The case of Fentalle Woreda, East Shewa zone, Oromia National Regional State, Ethiopia. M.Sc. Thesis, Haramaya University, Ethiopia.

Branson, R.E. and D.G. Norvell, 1983. Introduction to Agricultural Marketing. McGraw Hill, New York, USA., ISBN-13: 9780070072411, Pages: 521.

CSA., 2008. Production of major crops. Agricultural sample enumeration survey. Central Statistical Authority, Addis Ababa, Ethiopia.

Cunningham, III L.T., B.W. Brorsen, K.B. Anderson and E. Tostao, 2008. Gender differences in marketing styles. Agric. Econ., 38: 1-7.

Daniels, H. and S. Fors, 2015. Supply and value chain analysis of onions in Ethiopia. Department of Energy and Technology, Swedish University of Agricultural Sciences, Uppsala, Sweden.

Dessalegn, G., T.S. Jayne and J.D. Shaffer, 1998. Market structure, conduct and performance: Constraints on performance of Ethiopian grain markets. Working Paper 8, Grain Market Research Project, Ministry of Economic Development and Cooperation, Addis Ababa, January 1998.

FAO., 2005. Addressing marketing and processing constraints that inhibit agrifood exports: A guide for policy analysts and planners. FAO Agricultural Services Bulletin 160, Food and Agricultural Organization, Rome, Italy.

Hailu, A., 2016. Value chain analysis of vegetables: The case of Ejere district, West Shoa Zone, Oromia National Regional State of Ethiopia. M.Sc. Thesis, Haramaya University, Haramaya, Ethiopia.

IFAD., 2003. Rural micro-enterprise finance project. https://www.ifad.org/evaluation/reports/ppa/tags/philippines/505/1861155.

Makhura, M.T., 2001. Overcoming transaction costs barriers to market Participation of smallholder farmers in the Northern province of South Africa. Ph.D. Thesis, Department of Agricultural Economics, Extension and Rural Development, Faculty of Natural and Agricultural Sciences, University of Pretoria, South Africa.

Minot, N. and B. Sawyer, 2013. Agricultural production in Ethiopia: Results of the 2012 ATA baseline survey. International Food Policy Research Institute, Washington, DC., USA., April 21, 2013.

Pomeroy, R.S. and A.C. Trinidad, 1995. Industrial Organization and Market Analysis: Fish Marketing. In: Prices, Products and People: Analyzing Agricultural Markets in Developing Countries, Scott, G.J. (Ed.)., Lynne Rienner Publishers Inc., Boulder, Colorado.

Rabinowitch, H.D. and L. Currah, 2002. Allium Crop Science: Recent Advances. CABI Publishing, UK., ISBN-13: 9780851995106, Pages: 515.

Saccomandi, V., 1998. Agricultural Market Economics: A Neo-institutional Analysis of the Exchange, Circulation and Distribution of Agricultural Products. Van Gorcum Ltd., Assen, The Netherlands, ISBN-13: 9789023229391, Pages: 231.

Shiferaw, B., G. Obare and G. Muricho, 2006. Rural institutions and producer organizations in imperfect markets: Experiences from producer marketing groups in semi-arid Eastern Kenya. CAPRi Working Paper No. 60/October 2006, International Food Policy Research Institute, Washington, DC., USA.

Tadesse, A., 2011. Market chain analysis of fruits for Gomma Woreda, Jimma zone, Oromia national regional state. M.Sc. Thesis, Haramaya University, Haramaya, Ethiopia.

Takele, A., 2010. Analysis of rice profitability and marketing chain: The case of Fogera Woreda, South Gondar Zone, Amhara National Regional State, Ethiopia. M.Sc. Thesis, Haramaya University, Haramaya, Ethiopia.

Thakur, D.S., H. Lal, D.R. Thakur, K.D. Sharma and A.S. Saini, 1997. Market supply response and marketing problems of farmers in the hills. Indian J. Agric. Econ., 52: 139-150.

# Index